SOUNDSCAPE SEMIOTICS - LOCALISATION AND CATEGORISATION

Edited by **Hervé Glotin**

Soundscape Semiotics - Localisation and Categorisation
http://dx.doi.org/10.5772/45861
Edited by Hervé Glotin

Contributors

Chiung Yao Chen, Nozomu Hamada, Xiao-Li Zhong, Bo-Sun Xie, Tetsuya Takiguchi, Ryoichi Takashima, Yasuo Ariki, Zijian Liu, Lanbo Liu, Vasileios Exadaktylos, Mitchell Silva, Daniel Berckmans, Kim Fluitt, Tomasz Letowski, Pascale Giraudet, Olivier Dufour, Thierry Artieres, Hervé Glotin

Published by InTech

Janeza Trdine 9, 51000 Rijeka, Croatia

Publishing Process Manager Ana Pantar
Technical Editor InTech DTP team
Cover InTech Design team

First published February, 2014
Printed in Croatia

Additional hard copies can be obtained from orders@intechopen.com

Soundscape Semiotics - Localisation and Categorisation, Edited by Hervé Glotin
p. cm.
ISBN 978-953-51-1226-6

For EU product safety concerns:
IN TECH d.o.o., Prolaz Marije Krucifikse Kozulić 3, 51000 Rijeka, Croatia,
info@intechopen.com or visit our website at intechopen.com.

Meet the editor

Hervé Glotin is a professor at the Insitut Universitaire de France (since 2011) and Univ. of Toulon (since 2010), in the Systems & Information Sciences CNRS lab. He is leading the DYNI team on stochastic multimodal information retrieval. He received his master 1 in computer science from University Pierre et Marie Curie-Paris. He carried out his PhD at the Inst. of Perceptual Artificial Intelligence (IDIAP) CH and Inst. of Spoken Communication - Perception Team Grenoble on "Robust adaptive multi-stream automatic speech recognition using voicing and localization cues". In 2000 he has been involved as expert at Johns Hopkins CSLP lab with IBM human language team in audiovisual Large Vocabulary Speech Recognition. After two years as research engineer at CNRS lab on phonology and Semantic analyses, he has been an assistant professor position at University of Toulon in 2003. He then conducted researches on multimodal pattern analysis and retrieval systems, audiovisual indexing, cognitive models and bioacoustics. His research interests include signal processing, scene Understanding (vision, audition, bioacoustics), cognitive Systems and machine learning.

Contents

Preface

Objectives of this book

This book presents advanced acoustic analysis for complex scene analysis, understanding and monitoring soundscape, that is *soundscape semiotics*. The different chapters show that efforts are still to be produced to integrate robust processing and machine learning algorithms for scaled soundscape semiotics. They demonstrate the need for standardization within various fields like acoustic imaging, psycho-acoustics, computational auditory scene analysis and bioacoustics.

Acoustic mining processes play a major role in communication and exploration for most of the animals. They enable quick load and transfer of information tackling the reduced visibility and the long distances. When animals need to grasp a sound target in presence of competing signals, they may select specific features: a crucial process for their survival. For example, bats have a sensory apparatus very different from human being. They perceive the world primarily by sonar (echolocation), detecting the reflections from objects, of their own subtly modulated, high-frequency shrieks. Their brain correlates the outgoing impulses with the subsequent echoes. It enables bats to make precise localization and categorization from acoustics, estimation of distance, size, shape, motion, and texture of all kind of targets, obstacles, preys. It yields to comparable scene analysis to those human being get by vision. At much lower resolution, blind people are able to detect objects near them by a form of human-sonar, using vocal clicks or taps of a cane.

In the last decades, computational acoustic scene analysis was mostly dedicated to human audition, however bioacoustics is opening new paradigms in soundscape semiotics. Several methods are developed from mono or multiple microphones array, and are important in many applications, like for environmental conservation programs, and the international Scaled Acoustic Biodiversity Big Data Project SABIOD[1], that involves together major laboratories in signal processing, speech processing, machine learning and bioacoustics. The aim of this book is to demonstrate methods for soundscape semiotics, a new science, to detect and extract characteristics from specific sounds, to localize their source, and if possible, to reveal neuro-physiological cues.

1 SABIOD Scaled Acoustic Biodiversity Project : http://sabiod.univ-tln.fr

Worldwide challenges in acoustic categorization and localization

Recently, in 2013, three international challenges initiated objective evaluation of automatic separation and categorization of complex natural sounds: the calls of different bird species (that is the 'cocktail party' paradigm): the ICML4B[2] and NIPS4B[3] workshop challenges[4] [1, 2], and the MLSP 2013 challenge. In 2014, a bigger bird supervised categorization challenge (LifeClef) runs on 500 species of the amazon forest[5] over 14k records. Because the target are not always know *a priori,* due to the considerable amount of data generated by simple and efficient analog recorders, unsupervised soundscape semiotics methods have an increasing importance and are central into the 2014 ICML workshop 'Unsupervised Learning from Bioacoustic Big Data' (ICML uLearnBio[6]).

Other needs of soundscape semiotics concern sub-marine, like marine mammals monitoring using passive bioacoustics. One the most demonstrative experiment was presented during the 2005 DCL workshop [3] from five ocean bottom mounted (-1500m) widely spaced hydrophones (400 m distant) from NATO. An efficient algorithm [4, 5, 6, 7, 8] is based on the principle of transitivity of Time Delay of Arrival (TDOA) computed from correlation of each couple of hydrophones. It results in a high precision track without false alarm, robust to multiple sources [5, 6, 7, 8] illustrated in fig. 1, 2, 3.

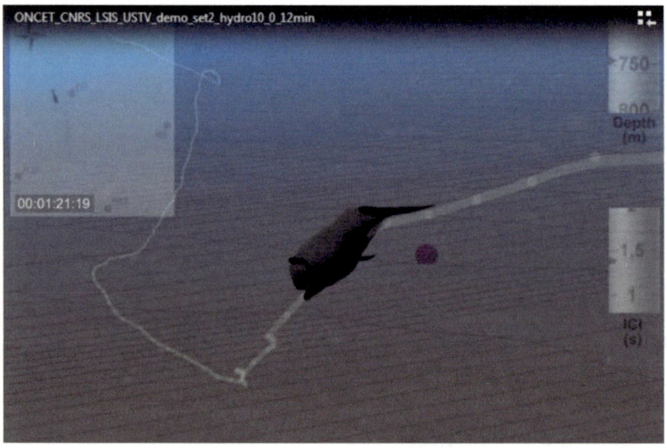

Figure 1. Video of precise 3D whale tracking by passive acoustics method, processed on the 2005 DCL challenge, NATO Bahamas recordings [4,6,8]. This video shows, without post process, the accurate estimations of depth and animal behavior in 4D (space and time). Available online video : http://sabiod.univ-tln.fr/tv [we thank P. Cosentino and L. Hauc for their help in this representation].

---

2 http://sabiod.univ-tln.fr/icml2013 [2]

3 http://sabiod.univ-tln.fr/nips4b [1]

4 Data of these challenges are still available for researches, see also Dufour et al. in this book or [1,2]

5 http://www.imageclef.org/2014/lifeclef/bird

6 http://sabiod.univ-tln.fr/ulearnbio/

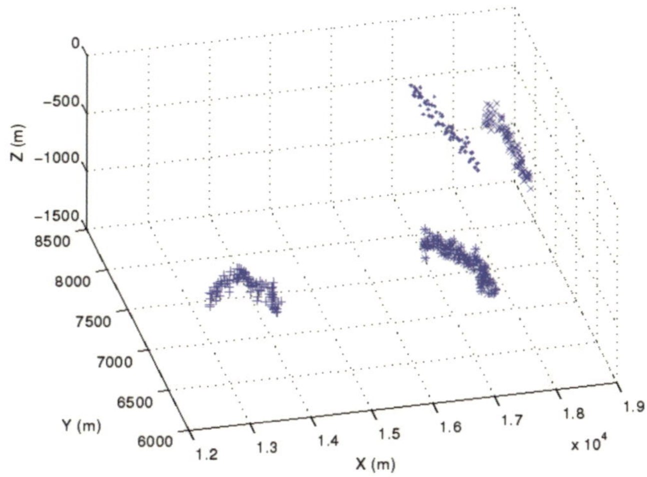

Figure 2. *Multiple Whale 3D tracking by passive acoustics, during half an hour, computed* [5,6,8] *from Bahamas data set 2003. Each symbol corresponds to 1 of the 4 whales diving in the area. They are clearly and continuously tracked showing correlated behaviors. Video at* http://sabiod.org/tv

Localization passive acoustics methods allow also to detect and track events on short base hydrophone array, as shown on 2 meters Nemo Onde array (INFN, CIBRA and the NEMO collaboration group). Some of these recordings were distributed in the 2009 DCL challenge. Taking into account the problem of data association, called the Rao-Blackwellized Monte Carlo data association (RBMCDA), a method allows to locate several sperm whales with a reasonable accuracy [7] illustrated Fig 3.

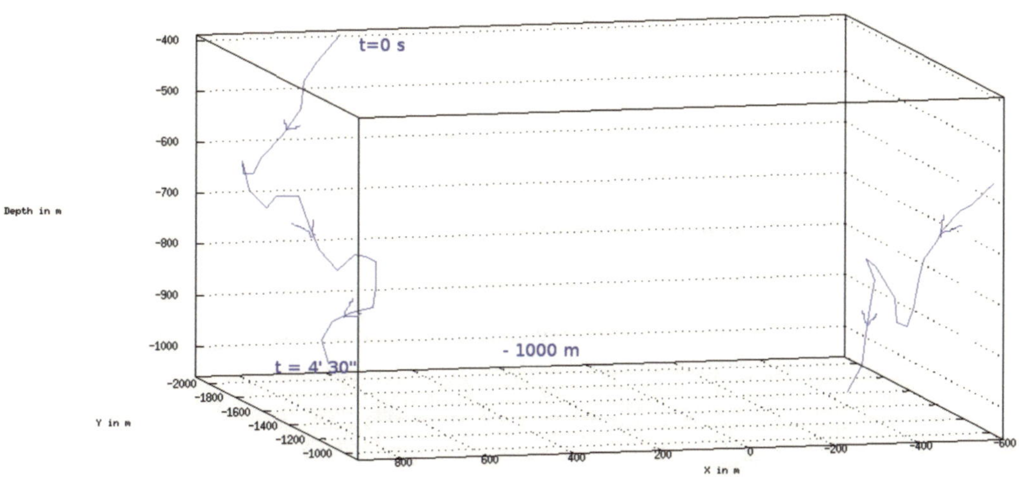

Figure 3. *Whale 3D tracking results from ETNA area recorded in 2005 on NEMO 2m short base hydrophone array. We see clearly 2 Physeter catodon diving together from -400m to -1000 m in 5 minutes* [6,7]. [We thank G. Pavan & Ricobene for NEMO recordings from INFN & CIBRA].

Perspectives

The source localization can then be joined to acoustic categorization [9], in several applications, as in speech processing. One of the first cocktail party speech recognizer was based on this paradigm [10]. The book is presenting most advanced cognitive based models for speech and/or soundscape analyses.

We have shown some examples of the soundscape semiotics challenges, within a general framework merging signal processing, pattern recognition and machine learning. Many of these activities are taking place into projects which aims to detect, cluster, classify and index acoustic big data in various ecosystems, at different space and time scales, in order to reveal information on the health of an ecosystem, yielding to new biodiversity insights. This scaled acoustic data science is a novel challenge that requires new methods.

Book organization

The chapters of this book assemble together auditory scene analysis, bioacoustics and human auditory system models. They have been selected for their high level and originality. The book is divided in two sections. First section - Advanced Signal Processing Methodologies for Soundscape Analysis contains 5 chapters -, and second section - Human Hearing Estimations and Cognitive Soundscape Analysis 3 chapters. The target audience comprises scholars and specialists in the field.

Acknowledgments

We thank each chapter author(s) of this collaborative project for their interest in grouping together different fields toward the same paradigm of soundscape semiotics.

I thank the interdisciplinary mission of the French National Research Center (MI CNRS), which supports the Scaled Acoustic Biodiversity SABIOD project (2012-2015...) http://sabiod.univ-tln.fr. I thank the Institut Universitaire de France (IUF) which supports my Chair on complex scene analysis (2011-2016), and the Université de Toulon and UMR CNRS LSIS for support of the computational auditory scene analysis and bioacoustic researches conducted in Dyni team.

I also thank my colleagues for their fascinating and motivating discussions on this interdisciplinary research field: O. Adam, G. Pavan, W. Zimmer, J. Sueur, Y. LeCun, S. Mallat, T. Artières, J.-P. Haton, P. Giraudet, S. Paris, J. Razik, F. Chamroukhi, L. Kindermann, J.-L. Schwartz, R. André-Obrecht, L. Besacier, C. Clark, P. Dugan, E.-M. Nosal, P. White, X. Halkias, A. Mishchenko, C. Laplanche, Y. Doh, J.-M. Prévot, D. Mauuary, F. Schnoller, J. Patris, F. Malige, P. Cosentino, F. Bénard, and other Postdocs, PhD students and colleagues.

Herve' GLOTIN, Pr.

Address all correspondence to: http://glotin.univ-tln.fr, glotin@univ-tln.fr

Institut Univ. de France (IUF), Paris & Univ. Toulon (UTLN),
Head of information DYNamics & Integration project (DYNI, UTLN & CNRS LSIS),
Head of Scaled Acoustic Biodiversity SABIOD.org CNRS project,

(a) Université de Toulon, 83957 La Garde Cedex, France,
(b) Institut Universitaire de France, Boulevard St Michel, 75006 Paris, France,
(c) Laboratoire des Sciences de l'Information & des Systèmes, CNRS UMR 7296,
(d) Aix Marseille Université, LSIS, ENSAM, AMU, Marseille, France

Cited References

[1] Glotin H., LeCun Y., Artières T., Mallat S., Tchernichovski O., Halkias X., Proc. of Neural Information Processing Scaled for Bioacoustics: from Neurons to Big Data, joint to NIPS Conference, ISSN 979-10-90821-04-0, http//sabiob.org/NIPS4B2013_book.pdf, 2013

[2] Glotin H., Clark C., LeCun Y., Dugan P., Halkias X., Sueur J., Proc. of the 1st workshop on Machine Learning for Bioacoustics, joint to ICML conference, Atlanta, ISSN 979-10-90821-02-6, http://sabiod.org/ICML4B2013_book.pdf, 2013

[3] Adam O., Samaran F. Detection, Classification and Localization of Marine Mammals using passive acoustics. 2003-2013: 10 years of internat. research, Dirac Ed., 2013

[4] Giraudet P., Glotin H., Real-time 3D tracking of whales by echo-robust precise TDOA estimates with a widely-spaced hydrophone array. Int. J. Applied Acoustics, Elsevier Ed., V67, Issues 11-12, pp 1106-1117, nov. 2006

[5] Glotin H., Caudal F., Giraudet P., Whales cocktail party: a real-time tracking of multiple whales, in Int. J. Canadian Acoustics, V36, p139-145, ISSN 0711-6659, 2008

[6] Glotin H., Giraudet P., Caudal F., Patent, Real time multiple whale tracking by passive acoustics, FR 2007 07/06162, Europe 2011, USA 2013

[7] Bénard F., Glotin H., Giraudet P., Whale 3D monitoring using astrophysic NEMO ONDE two meters wide platform with state optimal filtering by Rao-Blackwell Monte Carlo data association, in J. of Applied Acoustics, V71, pp. 994-999, 2010

[8] Bénard F., Glotin H, Giraudet P., Highly defined whale group tracking by passive acoustic Stochastic Matched Filter, Intech Ed., Advances in Sound Localization, ISBN 978-953-307-224-1, http://www.intechopen.com/articles/show/title/highly-defined-whale-group-tracking-by-passive-acoustic-stochastic-matched-filter, 2011

[9] Halkias X, Paris S. Glotin H., Classification of Mysticete sounds using machine learning techniques, in J. of the Acoustical Society of America, V134 (5), pp. 3496, nov. 2013

[10] E Tessier, F Berthommier, H Glotin, S Choi, A CASA front-end using the localisation cue for segregation and then cocktail-party speech recognition; Proc. IEEE ICSP, 1999

Section 1

Advanced Signal Processing Methodologies for Soundscape Analysis

Chapter 1

Source Separation and DOA Estimation for Underdetermined Auditory Scene

Nozomu Hamada and Ning Ding

Additional information is available at the end of the chapter

http://dx.doi.org/10.5772/56013

1. Introduction

In human-machine communication the separation of a target speech signal and localization of it in noisy environments are very important tasks. [1] For carrying out these tasks recent advanced sensor array signal processing is promising technology. [2] It utilizes the collection of multi-channel acoustic data by an array of microphones for detecting and producing output signals which is much more intelligible and suitable for communication and automatic speech recognition. [3]

BSS problem

Blind source separation (BSS) aims to estimate source signals by only using their mixed signals without any a priori information about mixing process and acoustic circumstances. The cocktail-party problem is one of the typical BSS problems. [1] Basically, the BSS problem can be solved by exploiting intrinsic properties of speech signals. Depending on the inherent properties there have been proposed lots of methods for BSS problems on speech signals. Among them the most widely applied approaches are the following two.

1. Independent component analysis (ICA)[4]-[8], and
2. Time-Frequency sparseness of source signals [9]-[14].

The ICA-based separation relies on *statistical independence* of speech signals in time-domain [5] [7] as well as in frequency-domain [6]. In addition, [8] proposed a dynamic recurrent separation system by exploiting the spatial independence of located sources as well as temporal dependence. On the other hand the second approach exploits the *sparseness* of speech signals in time-frequency (T-F) domain where only small number of T-F components are dominant in representing a speech signal. The T-F sparseness leads the disjoint property of T-F domain components, called W-disjoint orthogonality (WDO) property [11] [12], between speech signals. It means that at most one source dominates at every T-F points, in another word; different speech signals rarely generate the same frequency at the same time.

Though ICA approach performs well even in a reverberant condition, it is difficult to solve the underdetermined case in which the number of sources is greater than the number of sensors. Additionally, the frequency-domain ICA [6] the permutation ambiguity of its solution is a serious problem. It needs to align the separated frequency components that originate from the same source.

The T-F masking method which is the most popular sparseness-based approach is the topic concerned in this chapter. The representative method is known as DUET (Degenerate Unmixing Estimation Technique) [11]. A flow of conventional sparseness-based separation can be summarized as follows.

Sparseness-based T-F masking

Observed signals in T-F domain:

Transform time domain acoustic observations during few seconds to the T-F domain signals by applying short time Fourier transform (STFT) where a sparse representation of speech signal is obtained. [15] Thus the T-F components of a speech signal distribute in T-F domain without overlapping with T-F components of other speech signals.

Features of T-F cells:

As known in auditory scene analysis interaural time differences and level differences are significant spatial features of sources. [1] These localization cues are estimated from the differences in the direction and the distance of speakers. Actually, in microphone array the geometric parameters of sources can be obtained from phase differences and attenuation ratios at the mixture T-F cells.

Clustering T-F cells:

Under the WDO assumption the distribution of feature vectors obtained at all T-F cells makes as many clusters as the number of sources. The essential task of separation therefore turns out to cluster the feature vectors. The preliminary clustering method adopted in [9] - [12] is to make the histogram of features and to find the peaks corresponding the sources. Each T-F cell in the mixed signal is thereby associated with one peak depending on the distance in the cell's feature space.

Masking T-F cells:

Utilizing the clustering results individual binary masks are applied to the T-F domain spectrogram to detect the components that originate from individual sources.

Inverse transform:

A set of masked T-F components are inversely transformed by STFT and then it provides restored speech signal.

Remarks:

1. T-F domain sparseness in speech signals is also employed as a separation principle in the context of single channel or monaural signal source separation problem where harmonic structure in spectrogram is crucial for segregation.[16] [17]

2. Associated with the features of T-F cells conventionally used features are summarized in [13] and the features are evaluated from the separation performance point of view.

3. Clustering scheme in T-F masking would be crucial for high separation ability. Subsequent studies after DUET-like approaches [11][12], maximum-likelihood (ML) based method for real-time operation [18], k-means algorithm or hierarchical clustering, and EM algorithm [19] have been proposed. The method called MENUET [13] applies k-means algorithm to a vector space consisting of the signal level ratio and the frequency-normalized phase difference with appropriately weighting terms for effective clustering. They solve the optimization problem by adopting an efficient iterative update algorithm. In [14] k-means algorithm is applied clustering spatial features for arbitrary sensor array configuration even with wider sensor distance where spatial aliasing may occur. Their clustering procedure is divided into two steps, the first one of which is applicable to the non-aliasing or lower frequency band and the second one treats the remaining aliasing occurred frequency band.

DOA estimation

Localization of acoustic sources using microphone array system is a significant issue in many practical applications such as hands-free phone, camera control in video conference system, robot audition, and so on. The latter half of this chapter focuses on the Direction-Of-Arrival (DOA) estimation of sources. Since this monograph interests in speech signals, we make no mention of the methods addressed for narrow-band signals, for instance in radar/sonar processing. There have been proposed a large number of DOA estimation methods for broadband signals [20], [21]. Typical array processing approaches are;

1. Generalized Cross-Correlation (GCC) methods [22]

2. Subspace approaches using spatial covariance matrix of observed signals [23]

3. T-F domain sparseness-based approaches [11],[24]-[27]

4. ICA separation based approaches [28]

The first category of GCC method is to estimate the delay time that maximizes a generalized cross-correlation function between the filtered outputs of the acquired signals at microphones. The phase transform (PHAT) method [22] exploits the fact that the Time-Delay-Of-Arrival (TDOA) information is conveyed in the phase. Although GCC methods are usually performed well and are also computationally efficient for single source case, it does not cope with multiple sources case in which this chapter interests.

The second category is the subspace analysis applying a narrowband signal model. The analysis uses the properties in the spatial covariance matrix of multichannel array observations. The MUSIC-like algorithms are well-known methods for narrowband target signals. For broadband signals such as speech, several frequency-domain approaches have been proposed. The subspace-based approaches for small number of sensors have to overcome two drawbacks, one of which is the limited precision for DOA estimation, and the other is that it is unable to deal with the underdetermined case.

Sparseness-based approaches

The third category of the DOA estimation algorithms is based on sparseness of speech signals and is closely related to the BSS. Source sparseness assumption implies WDO or its weaker condition TIFROM [24]. These conditions are the crucial properties to solve DOA problems for underdetermined multiple sources. The BSS approach associated with these assumptions is a group of T- F masking framework. In DUET-like methods [9]-[14], the delay time or the frequency-normalized ratio of the frequency-domain observations at each T-F point is used to compute the TDOA. An alternative DOA estimation method proposed by Araki et al. [27], in the context of k-means algorithm, estimates DOA as the individual centroid of each cluster of normalized observation vectors corresponding to an individual source. The DEMIX [25] algorithm introduces a statistical model in order to exploit a local confidence measure to detect the regions where robust mixing information is available. The computational cost of DEMIX would be high due to performing the principal component analysis for every local scatter plot of observation vectors at individual T-F points.

For addressing robust cocktail-party speech recognitions the localization cue such as TDOA or spatial direction evaluated at each T-F cell has a central role. As in [29][30], integrating approaches the segregation/localization of sound sources and speech recognition against background interferences are significant CASA (Computational Auditory Scene Analysis) front-ends.

DOA Tracking

Not only estimating but also tracking sound sources draws lots of attentions recently in robot auditory systems. For instance, speaker's DOA tracking by microphone array mounted on mobile robot is the problem of moving sources and moving sensors.

BSS and DOA Problems:

The underlying BSS and DOA estimation problems addressed in this chapter are listed as follows:

a. Use of a pair of microphones

b. Multiple simultaneously uttered speech signals under the assumption that the number of sources is known a priori

c. Underdetermined cases, where the sources outnumber the sensors

d. The inter-sensor distance is bounded so as to avoid spatial aliasing (for instance, less than 4 cm spacing for an 8 kHz sampling rate)

While stereophonic sensor is the simplest sensor array, the study of how to improve the separation performance and to obtain accurate DOA by a pair of microphones is meaningful because any complex array configuration can be considered as an integration of these.

The rest of this chapter is organized as follows. In section 2, problems of underlying BSS and DOA estimation are described in detail. The proposed BSS method based on a frame-wise scheme is introduced in section 3. Section 4 describes a DOA estimation algorithm by using T-F cell selection and the kernel density estimator. The last section concludes this chapter.

2. Problem descriptions

2.1. Observation model

Source mixing models in time domain and its T–F domain description are described as follows. All discrete time signals are sampled version of analog signals with sampling frequency f_S. Suppose N source signals $s_1(t)$, $s_2(t)$, $\cdots$, $s_N(t)$ are mixed by time-invariant convolution and the observed signals $x_1(t)$, $x_2(t)$, $\cdots$, $x_M(t)$ at M sensors with omni-directive characteristic are described as:

$$x_m(t) = \sum_{i=1}^{N} \sum_{\tau} h_{mi}(\tau) s_i(t-\tau), \tag{1}$$

where $h_{mi}(\tau)$ represents the impulse response from i-th source to m-th sensor. Observed signals $x_m(t)$ (m=1~M) are converted into T–F domain signals $X_m[k, l]$ by using L-point windowed STFT as written by

$$X_m[k,l] = \sum_{r=-L/2}^{L/2-1} x_m(r+kS) win(r) e^{-j\frac{2\pi l}{L} r}, \qquad k = 0 \sim K, l = 0 \sim \frac{L}{2} \tag{2}$$

where r is dummy variable in convolution sum operation, $win(r)$ is a window and S is the window shift length. Here, we apply half window size overlapping transformation, namely $S = L/2$ in (2). Transformed T–F mixture model of Eq.(1) can be described by the instantaneous mixtures at each time frame index k and frequency bin l.

$$X_m[k,l] = \sum_{i=1}^{N} H_{mi}[l] S_i[k,l] \tag{3}$$

where $H_{mi}[l]$ is the frequency response (DFT) of $h_{mi}(t)$, $S_i[k, l]$ is the windowed STFT representation of i-th source signal $s_i(t)$, and the point $[k, l]$ is called *"T-F cell"* in this chapter.

Assuming an anechoic mixing, the source signals which we want to recover are alternatively redefined as the observed signals at the first mixture $x_1[k, l]$. In this case, the following mixing models in the T–F domain are henceforth considered without loss of generality.

$$\begin{aligned} X_1[k,l] &= \sum_{i=1}^{N} S_i[k,l], && \text{a} \\ X_m[k,l] &= \sum_{i=1}^{N} H_{mi}[l] S_i[k,l] \qquad m = 2 \sim M && \text{b} \end{aligned} \tag{4}$$

where $S_i[k, l]$ and $H_{mi}[l]$ are different from $\mathsf{S}_i[k, l]$ and $\mathcal{H}_{mi}[l]$ in (3), $S_i[k, l]$ is the i-th source signal observed at the first sensor (m=1), and $H_{mi}[l]$ eventually represents the DFT domain operation of the transfer function with relative attention and delay between m-th and the first sensors.

From then on, consider the mixture of two sources $S_1[k, l]$ and $S_2[k, l]$ which are received at a pair of microphones. Their mixture system (4a) and (4b) can thus be expressed as

$$\begin{bmatrix} X_1[k,l] \\ X_2[k,l] \end{bmatrix} = \begin{bmatrix} 1, & 1 \\ H_{21}[l], & H_{22}[l] \end{bmatrix} \begin{bmatrix} S_1[k,l] \\ S_2[k,l] \end{bmatrix} \tag{5}$$

2.2. Basic assumptions

As stated in Section 1, the WDO is commonly supposed in sparseness-based separation approaches. At first, we denote the T-F domain Ω on which $S_1[k, l]$ and $S_2[k, l]$ are defined

$$\Omega := \{[k,l], k = 0 \sim K, l \in B\} \tag{6}$$

where $B := [l_1, L/2]$ is the frequency band after deleting lower frequency components which do not exist in actual speech signals, and $l_1 = \lfloor f_1 L / f_s \rfloor$ means the Gauss floor function, and f_1 is the analog lowest frequency of speech components such as 80Hz in later experiments.

Next, define the T-F supports $\Omega_i (i=1, 2)$ of $S_i[k, l] (i=1, 2)$ by

$$\Omega_i := \{[k,l] \,\|S_i[k,l]\| > \varepsilon\} \quad i = 1,2 \tag{7}$$

where $\varepsilon(>0)$ is a sufficiently small value. Although, in theory, the support of $S_i[k, l] (i=1, 2)$ is defined by the condition $|S_i[k, l]| \neq 0$, Eq. (7) gives a set of components of actual signals except noise-like ones satisfying $|S_i[k, l]| < \varepsilon$.

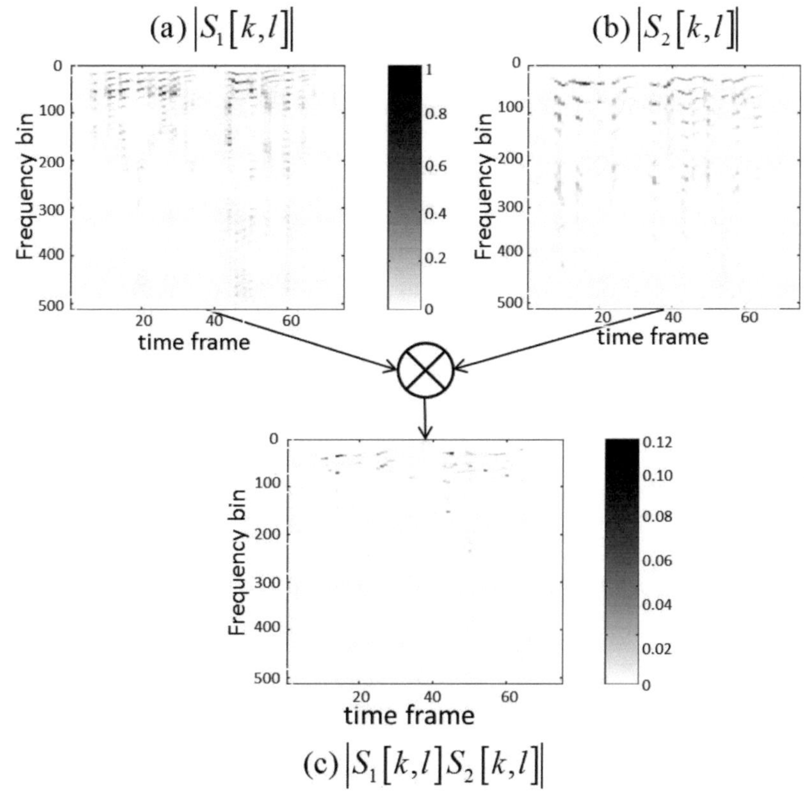

Figure 1. WDO property of two real speech signals

We may consequently express the WDO assumption between two source signals $s_1(t)$ and $s_2(t)$ by the disjoint condition

$$\Omega_1 \cap \Omega_2 = \phi(empty\ set) \tag{8}$$

This can equivalently be represented as follow.

$$S_1[k,l]S_2[k,l] = 0 \qquad \text{at any } [k,l] \tag{9}$$

The verification of above WDO condition for actual speech signals is performed in Fig. 1 where (a) and (b) show spectrograms of two speech signals in the T-F domain, and (c) shows their multiplication in which we see rarely overlapping between two spectrograms.

Obviously, the supports of $X_1[k, l]$ and $X_2[k, l]$ are coincident and it, denoted by Ω_X, can be given as

$$\Omega_X = \Omega_1 \cup \Omega_2 \tag{10}$$

In addition the following null component domain, denoted by Ω_N, is also introduced as

$$\Omega_N = \bar{\Omega}_X = \overline{\Omega_1 \cup \Omega_2} \qquad \bar{X}: \text{complementary set of } X \tag{11}$$

The WDO condition (8) accordingly derives that the T–F domain representation of the mixed signal $X_1[k, l]$, given by Eq.(5), can be decomposed into the following three parts with no overlapping in Ω.

$$X_1[k,l] = \begin{cases} S_1[k,l] & [k,l] \in \Omega_1 \\ S_2[k,l] & [k,l] \in \Omega_2 \\ 0 & [k,l] \in \Omega_N \end{cases} \tag{12}$$

2.3. Source separation

Under the WDO assumption expressed in (12), the binary masking in the T–F domain is performed as follow:

Clustering the T-F cells in the support Ω_X of the mixture $X_1[k, l]$ into two sub-regions Ω_1 and Ω_2, the separated source estimates in T-F domain, $\hat{S}_1[k, l]$ and $\hat{S}_2[k, l]$, are obtained by applying the masks

$$M_i[k,l] = \begin{cases} 1 & [k,l] \in \Omega_i \\ 0 & \text{otherwise.} \end{cases} \quad (i = 1,2) \tag{13}$$

on $X_1[k, l]$ as follows.

$$\hat{S}_i[k,l] = M_i[k,l] X_1[k,l] \quad (i = 1,2) \tag{14}$$

Clustering features

The separation task is to classify T–F cells composing the support Ω_X of $X_1[k, l]$ into either Ω_1 or Ω_2. A pair of $X_1[k, l]$ and $X_2[k, l]$ is used to characterize a T-F cell $[k, l]$ at which spatial features are introduced, and the clustering process is performed in the estimated feature space.

Effective features must be the signal level or attenuation ratio defined by

$$\alpha[k,l] := \left|\frac{X_1[k,l]}{X_2[k,l]}\right| \tag{15}$$

and the arrival time difference defined by the frequency-normalized phase difference (PD) between $X_1[k, l]$ and $X_2[k, l]$ as

$$\delta[k,l] := \frac{L}{2\pi f_s l}\phi[k,l] \tag{16}$$

where $\phi[k, l]$ is the PD as defined by

$$\phi[k,l] = \angle X_1[k,l] - \angle X_2[k,l] \tag{17}$$

Other features used for characterizing T-F cells are listed in [13]as well as the attenuation ratio modifications. It is noted that the attenuation ratio would not give distinctive difference for short distance microphone array. In our experimental setup, for example, the distance between microphones is 4cm in order to avoid spatial aliasing at 8kHz sampling rate.

Clustering scheme

For given features at T-F cells in Ω_X, clustering of these is the next step. In DUET where a pair of microphones is used, the two dimensional histogram of feature vectors $\{\alpha[k, l], \delta[k, l]\}^T$ within a time interval, such as for several seconds, is generated and the clustering is performed by finding the maximum peaks which are corresponding to sources. When the attenuation feature is omitted the clustering problem is solely performed based on time delay histogram distribution. The dimension of feature space will be higher for array configuration with many microphones than two. For these cases more sophisticated clustering scheme such as k-means algorithm or EM algorithm [19] should be adopted.

Inverse STFT

The final stage of the separation process is to obtain time domain separated signals $\hat{s}_i(t)$ $(i=1, 2)$ by applying the inverse STFT.

3. Sound source separation

3.1. Phase–difference vs. frequency data

As a T-F cell's feature depending on the spatial location difference of sources, our strategies exploit a frame-wise, namely, a time sequence of phase difference of observations versus frequency (PD-F) distribution. In a k-th frame, the point plot of the PD-F is defined as a collection of two-dimensional vectors at k-th frame $p_k(l)$as

$$\boldsymbol{p}_k(l) := \{l, \phi[k,l]\}^T, \quad l \in B \quad k \in [1,K] \tag{18}$$

An example of PD-F in $(l,\ \phi)$-plane and its time sequence for the mixture of two speech signals are shown in Fig.2 (a) and (b) respectively.

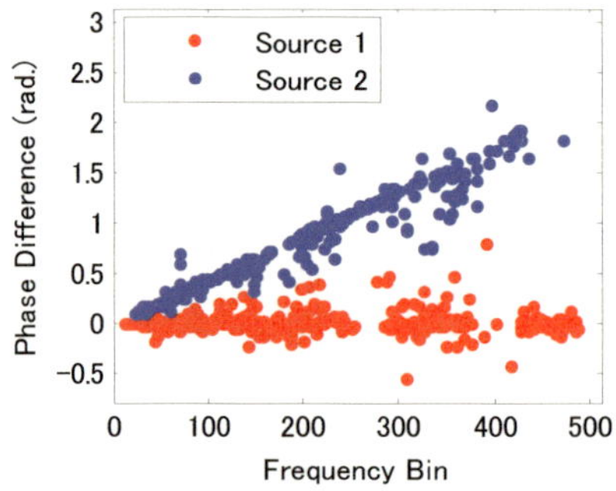

(a) PD-F plot

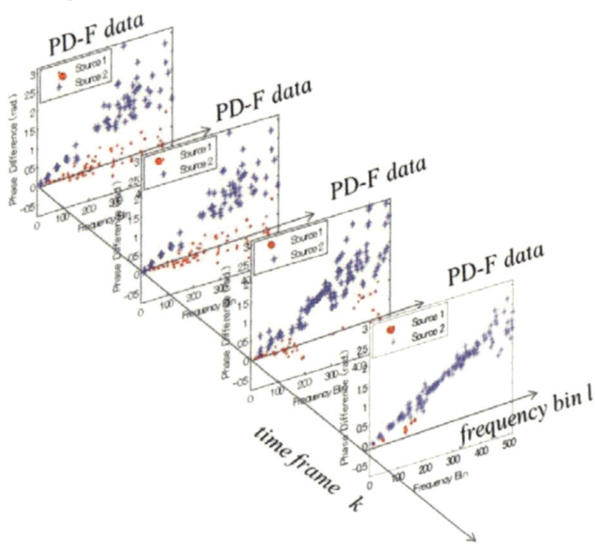

Figure 2. PD-F and time sequence of PD-F (Blue and red points respectively corrrespond invividual source components)

The relationship between the gradient β of a vector in PD-F plane defined in Eq.(18) and the source direction θ is: [33]

$$\beta = \left(\frac{2\pi}{L}\right)\frac{f_s d}{c}\sin\theta \tag{19}$$

where d is the distance between the sensors, c is the sound velocity, and θ is the direction of source. Here $\theta=0$ corresponds to the broadside direction and the term $(d/c)\sin\theta$ represents the wave arriving delay between microphones. For example, the dot distribution in Fig.2 (a) concentrates along two lines corresponding to two source directions. By determining the gradients of these lines two directions of sources are estimated from the relationship of (19).

The conventionally utilized features associating with delay time at each T-F cell can be estimated from the frequency normalization of PD-F dot corresponding to individual T-F cells. Unlike the conventional delay-like features PD-F dots keep a linear dot distribution on the plane and it is effectively utilized in both following source separation and direction finding methods.

3.2. Frame categorization

Fig. 3(a) shows two simultaneously uttered speech signals. In the figure four frame time points k_1 - k_4 indicated by the red rectangular parts are shown as the following four types of source signal activity states:

Frame k=k_1; No source signal is active (Non Source Active:NSA)

Frame k=k_2; Only the first source is active (Single Source SSA)

Frame k=k_3; Only the second source is active (Single Source SSA)

Frame k=k_4; Both sources are active (Double Source Active:DSA)

Here we may define three sets of time-frame indeces as follows:

The whole set of time-frames, denoted by **K**:={1,.....,K}, is categorized into three sets with no overlapping.

$$\boldsymbol{K} = \boldsymbol{K}_{NSA} \cup \boldsymbol{K}_{SSA} \cup \boldsymbol{K}_{DSA} \tag{20}$$

In addition, we define the following Sourse Active(SA) frame index set.

$$\boldsymbol{K}_{SA} = \boldsymbol{K}_{SSA} \cup \boldsymbol{K}_{DSA} \tag{21}$$

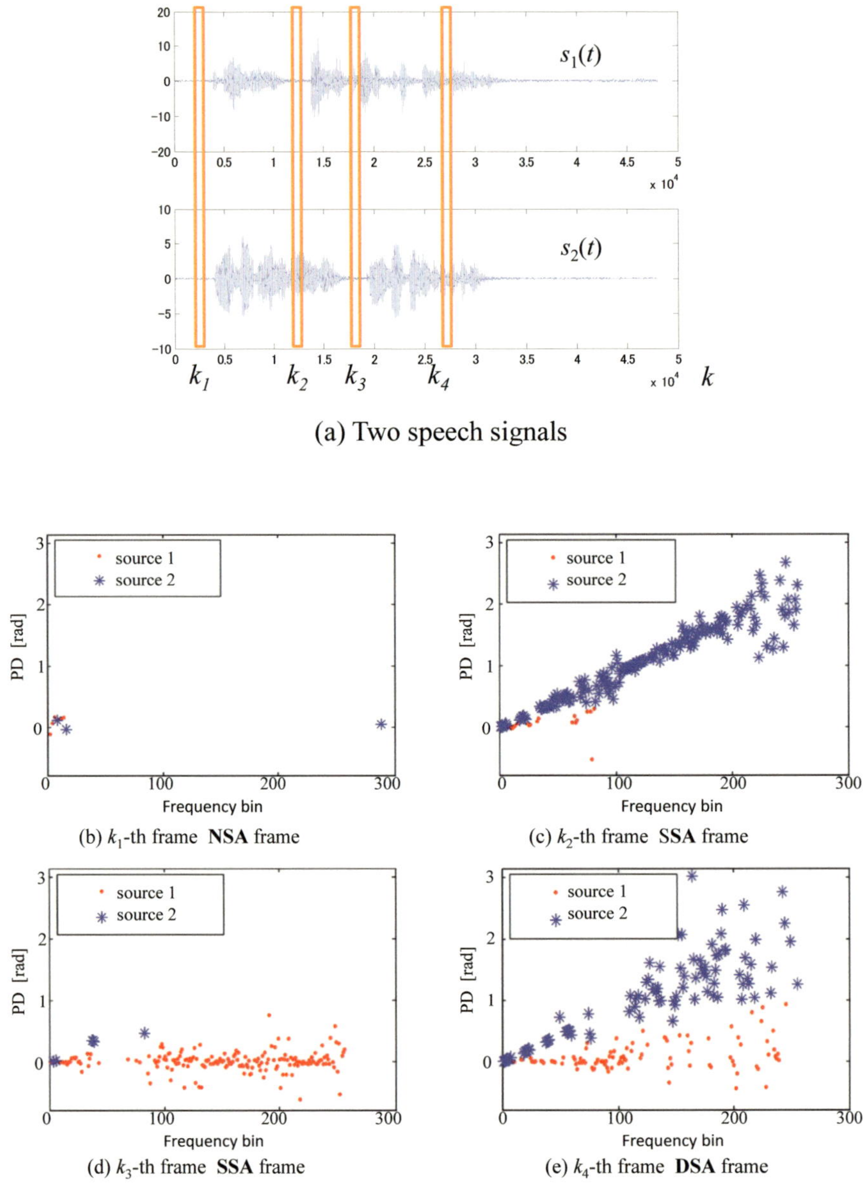

Figure 3. Frame categorization (NSA, SSA, DSA)

Above frame categorization suggests the source separation algorithm consisting of the following two parts:

- Assign each T-F component at SSA frame to either source by identifying the direction.
- Apply separation algorithm solely to DSA frames

The detail of these will be described in the next section.

http://dx.doi.org/10.5772/56013

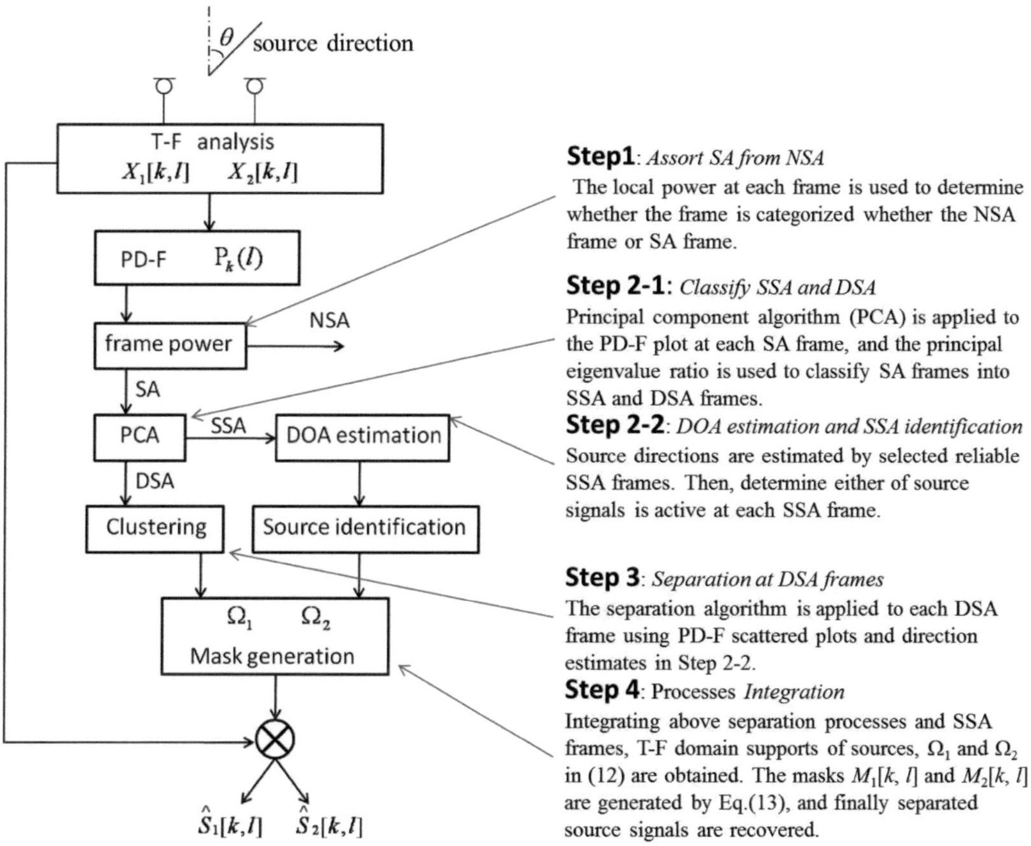

Figure 4. Flow of source separation method

3.3. Source separation algorithm

Outline of the method

The outline of the separation method using PD-F plot is shown in Fig.4 and summarized.

Step1: *Discriminate SA from NSA*

The following average power at a frame is employed to check the presence of speech signal at the frame.

$$E(k) := \frac{1}{L/2 - l_1 + 1} \sum_{l \in B} \left| X_1[k,l] \right|^2 \tag{22}$$

Here, the threshold operation is valid for basic voice activity detection as follow.

$$\boldsymbol{K}_{SA} = \{k | E(k) > Th_{SA}\} \tag{23}$$

where Th_{SA} is determined by a pre-experiment of noise level estimate during no utterance. In later experiments, we applied the following formula.

$$Th_{SA} = E_0 + 2\sigma_E \tag{24}$$

where E_0 is the average noise power estimate and σ_E is the standard deviation estimate given by respectively.

$$E_0 := \frac{1}{|\boldsymbol{k}_{NSA}|} \sum_{k \in \boldsymbol{k}_{NSA}} |X_1[k,l]|^2 \tag{25}$$

$$\sigma_E := \sqrt{\frac{1}{|\boldsymbol{k}_{NSA}|} \sum_{k \in \boldsymbol{k}_{NSA}} \left(E(k) - E_0\right)^2} \tag{26}$$

Step 2-1: *Classify SA into SSA and DSA*

At each $k \in K_{SA}$ PCA is applied to the set of vectors $p_k(l)$ by computing the following 2 × 2 covariance matrix.

$$\boldsymbol{R}_k := \frac{1}{L/2 - l_1 + 1} \sum_{l \in B} \boldsymbol{p}_k(l) \boldsymbol{p}_k^{\mathrm{T}}(l) = \begin{bmatrix} R_{11}(k) & R_{12}(k) \\ R_{21}(k) & R_{22}(k) \end{bmatrix}. \tag{27}$$

Denoting the eigenvalues of R_k by $\lambda_1(k)$ and $\lambda_2(k)$ (assume $\lambda_1(k) \geq \lambda_2(k)$), the ratio of the principal eigenvalues defined by

$$r(k) := \frac{\lambda_2(k)}{\lambda_1(k)}. \tag{28}$$

is introduced to discriminate the SSA frames from the DSA. As shown in Fig.3 (c), (d), PD-F vector distribution at a SSA frame tends to concentrate around the first principal axis. This observation leads to the following discrimination of SSA from DSA frames and the estimation of the source directions.

The following criterion is applied to detect SSA frames.

$$\boldsymbol{K}_{SSA} = \{k | r(k) < Th_{SSA}\} \tag{29}$$

where Th_{SSA} is determined experimentally.

Step 2-2:*DOA estimation and SSA identification*

Define the normalized eigenvector of the first principal eigenvalue as

$$\boldsymbol{e}_1(k) := \begin{bmatrix} \cos\beta(k) \\ \sin\beta(k) \end{bmatrix} \tag{30}$$

where $\beta(k)$ is the gradient of the principal axes at k-th frame. The histogram of the set

$$\{\beta(k), k \in \boldsymbol{K}_{SSA}\} \tag{31}$$

will have two peaks which are corresponding two source directions θ_1 and θ_2 caluculated by Eq. (19). By clustering the set of θ into two groups according to the distance from θ_1 and θ_2, each SSA frame in K_{SSA} is classified into each one of the sources from the direction θ_1 and θ_2.

Double Source Active (DSA)

For given set of DSA frames K_{DSA}, the clustering of the vectors $p_k(l),\ l \in B$ into two sets is the problem. Before describing this separation algorithm, three frequency bands, denoted by B_{high}, B_{low}, and B_{mid}, are introduced to use in the following separation algorithm.

Frequency Bands

The following three frequency bands are defined respectively.

$B_{high} := \{l \mid l_2 < l < L\ /2\}$, $B_{low} := \{l \mid l_1 < l < l_2\}$, $B_{mid} := \{l \mid l_2 < l < l_3\}$

where $l_i = \lfloor f_i L\ / f_s \rfloor$, $(i = 2, 3)$, f_2 is set 400Hz, and f_3 is set 1kHz in later experiments.

The idea of source separation at DSA frames utilizing these bands is divided into two parts according to above frequency bands.

1. The first scheme, called initial separation, is applied to the T-F cells in B_{high} based on the directions of sources which have been estimated at the SSA frames previously.

2. The clustering in B_{low} is performed utilizing a harmonic structure relationship between the spectral components in B_{low} and that of B_{mid}. The harmonic structure in B_{mid} can be obtained by the initial separation results in B_{high}.

1. *Initial separation*

Denote the source directions estimated in SSA frames by θ_1 and θ_2, and their corresponding gradients in PD-F plane are β_1 and β_2 as defined in Eq.(31). The points on these two lines can be expressed as

$$\phi[k,l] = \beta_i l \quad (i = 1,2) \tag{32}$$

At $k \in K_{DSA}$, the nearest neighbor rule gives the binary mask $\widetilde{M}_i[k, l]$ in B_{high} which is defined as

$$\tilde{M}_i[k,l] = \begin{cases} 1, & if\ i = \arg\min\limits_{c} \left| \phi[k,l] - \beta_c l \right|, l \in B_{high} \\ 0, & otherwise. \end{cases} \tag{33}$$

As a result, the separated individual signals $\widetilde{S}_i[k, l]$ (i = 1, 2) are represented by

$$\tilde{S}_i[k,l] = \tilde{M}_i[k,l] X_1[k,l], \quad l \in B_{high} \tag{34}$$

2. *Separation in B_{low}*

Local maximum points in B_{mid}

The final task for separation process is to generate individual mask applied to B_{low}. In this final separation process, the observed amplitude spectrum given by $|X_1[k, l]|$ with $l \in B_{low}$ is compared with the initially separated spectra $\widetilde{S}_1[k, l]$ and $\widetilde{S}_2[k, l]$ with $l \in B_{mid}$ in terms of harmonic relationships. At first, with the help of local maximum frequencies of $|\widetilde{S}_i[k, l]|$, harmonic structure in B_{mid} is estimated for each separation spectra. We denote the obtained local maximum frequencies of $|\widetilde{S}_i[k, l]|$ are $b_{i1}(k)$, $b_{i2}(k)$, $\cdot\cdot\cdot$, and the number of local maxima in B_{mid} is $q_i(k)$.

Harmonics estimation

The distance of adjacent harmonics $\Delta d_i(k)$ is defined as

$$\Delta d_i(k) = b_{i2}(k) - b_{i1}(k),\ q_i(k) > 2 \tag{35}$$

When $q_i(k) = 0$ or 1, we regard that there is no harmonic in the frame k. The estimated harmonics in low frequency band $g_{in}(k)$ is

$$g_{in}(k) = b_{l1}(k) - \Delta d_i(k) n, \tag{36}$$

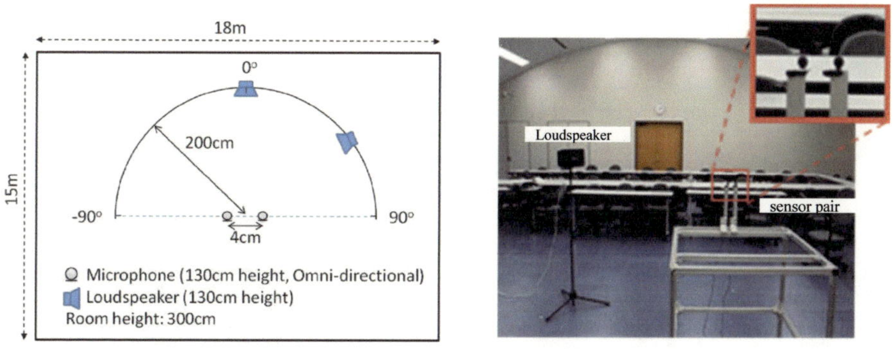

(a) Experimental environments (b) Experimental setup

Figure 5. Experiments

where $n=1, 2, 3, \cdots$, $g_{in}(k) \in B_{low}$, and $g_{in}(k)$ means the harmonic structure of source i at frame k.

Massk generation

We assume that the bandwidth of each harmonics is the same, and use 5 adjacent cells as bandwidth in T–F domain. The mask in B_{low} is defined

$$\overline{M}_i[k,l] = \begin{cases} 1, if\ g_{in}(k) - 2 < l < g_{in}(k) + 2,\ and \\ \qquad q_i(k) \geq 2, l \in B_{low}, n = 1,2,3,\cdots \\ 0, otherwise. \end{cases} \tag{37}$$

The integrated mask combining Eq. (33) and Eq. (37) is represented by

$$M_i[k,l] = \tilde{M}_i[k,l] + \overline{M}_i[k,l]. \tag{38}$$

Finally, the separated signals are obtained as shown in Eq.(14).

3.3. Experiments

Experimental condition

Some real life experiments are performed in a conference room to evaluate the separation methods. Fig.5(a),(b) show the experimental environments and the setup. The experimental parameters are show in Tab.1. One source was placed at the broadside (θ=0°) and the location of the other source is varied from 0° to 80° at intervals of every 10°.

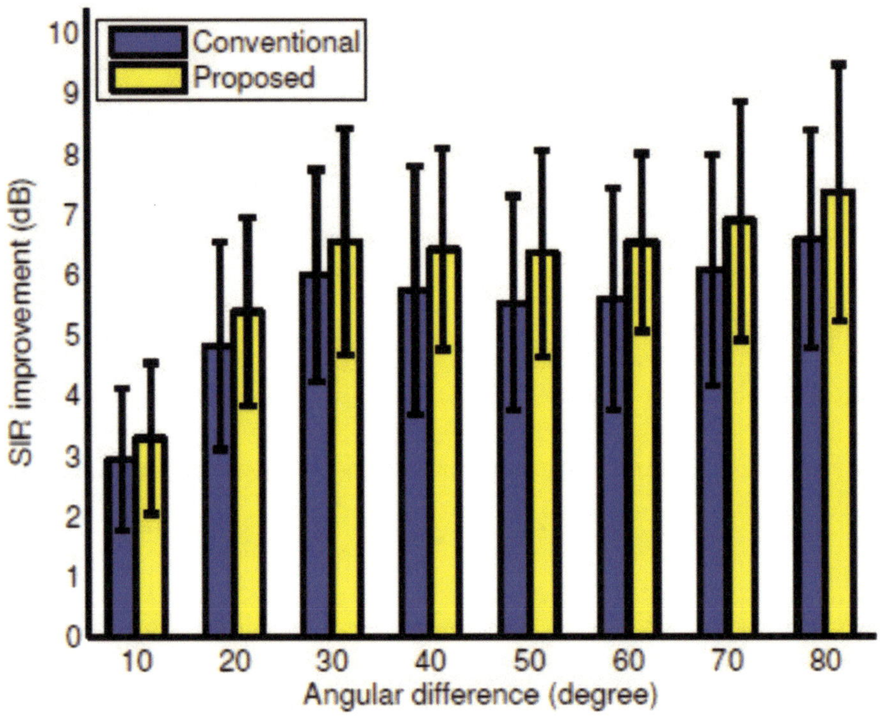

Figure 6. Experimental results (SIR)

Fig. 6 shows the average signal-to-interference ratio (SIR) improvement brought by the proposed and the conventional DUET method. The SIR improvement at the first sensor is defined as follows.

$$\text{SIR}_i \text{ improvement} = \text{Output SIR}_i - \text{Input SIR}_i \tag{39}$$

Where

$$\text{Input SIR}_i = 10\log_{10}\frac{\| s_i(t)\|}{\| s_j(t)\|}, \quad \text{Output SIR}_i = 10\log_{10}\frac{\| y_{ii}(t)\|}{\| y_{ij}(t)\|}$$

The proposed frame-wise PD-F approach exceeds the conventional method in terms of SIR improvement. The average improvement in our experiments is 6.22dB over the DUET. The most significant contribution in SIR improvement is made by the separation process in DSA frame which is 4.28dB. [31]

Source signal duration	5s speech signals
Sampling Frequency	8 kHz
Sound Velocity	340 m/s
Window	Hamming
STFT Frame Length	1024 sample
Frame Overlap	512 sample

Table 1. Experimental Parameters

4. DOA estimation

The DOA estimation method discussed in this section is based on the following three novel approaches.

1. Inspired by the ideas of TIme-Frequency Ratio Of Mixtures (TIFROM)-like assumptions, a novel reliability index is introduced. The selected cells with higher reliability are solely utilized for DOA estimation.
2. A statistical error propagation model relating PD-F and the consequent DOA is introduced. The model leads to a probability density function (PDF) of the DOA, and hence the DOA estimation problem is reduced to finding the most probable points of the PDF.
3. Source directions are determined using the kernel density estimator by utilizing the proposed bandwidth control strategy.

DOA information

Under the assumption of anechoic mixing with no-attenuation model and WDO in Eq. (5), the ratio between two observations $X_m[k, l]$ (m=1,2) is represented by

$$\frac{X_2[k,l]}{X_1[k,l]} = \frac{H_{2n}[l]}{H_{1n}[l]} = \exp\left[j\frac{2\pi f_s l}{L} \times \frac{d}{c} \sin\theta \right], \tag{40}$$

where θ is the direction of source which is dominant at [k,l]. The phase difference (PD) $\phi[k,l]$ between two observations $X_m[k, l]$ (m=1,2) defined by Eq. (17) is related to the angle θ as follows.

$$\phi[k,l] = \frac{2\pi f_s ld}{Lc} \sin\theta = \Delta\omega Tl \sin\theta, \tag{41}$$

where $T = d/c$ is the maximum delay time between sensors, and $\Delta\omega = 2\pi f_S / L$ is the unit frequency width in L-point STFT. From Eqs. (16) and (41), the TDOA normalized by T, denoted by $\tau[k, l]$, can be represented as follows.

$$\tau[k,l] = \sin\theta = \frac{\phi[k,l]}{T\Delta\omega l} \tag{42}$$

4.1. Reliable T–F cell selection

As stated in 2.2, the following selection processes are applied only to the T-F cells in the support Ω_X of $X_1[k, l]$ as in 2.2. This eventually reduces the computation time by eliminating noise-like T-F components.

Since the PD estimation by (17) is subjected to unavoidable error, the success of DOA estimation is generally expected if reliable PD data are selected to use and outliers are eliminated. Likewise in [24], the following assumption is employed. When a source is dominant in a set of cells, all delays in it will take almost the same value; hence, the delay (42) and obviously the PD data (17) in this set are expected to be reliable. Conventionally, the confidence measure is obtained from the results of applying the principal component analysis to a set of steering vectors in individual horizontal and vertical T-F regions. Unlike this approach, the normalized delays $\tau[k, l]$ given by Eq.(42) are used to evaluate the attribute consistency of the T-F cells. According to the above assumption, two types of T-F regions around a cell $[k, l]$ are considered: a temporal neighborhood $\Gamma_t[k, l]$ and a frequency neighborhood $\Gamma_f[k, l]$,

$$\Gamma_t[k,l] := \{[k+y,l] \,|\, |y| \le Y\}, \Gamma_f[k,l] := \{[k,l+z] \,|\, |z| \le Z\}, \tag{43}$$

where integers Y and Z determine the numbers of cells in these regions, as denoted by $|\Gamma_t[k, l]| := 2Y + 1$ and $|\Gamma_f[k, l]| := 2Z + 1$.

For each $\Gamma_t[k, l]$ and $\Gamma_f[k, l]$, the standard deviations of the normalized delays $\sigma_{\Gamma t}[k, l]$ and $\sigma_{\Gamma f}[k, l]$ are calculated by

$$\sigma_\Gamma[k,l] = \frac{1}{|\Gamma|} \sum_{[p,q]\in\Gamma} \left(\delta[p,q] - \mu_\Gamma[k,l]\right)^2 \tag{44}$$

$$\mu_\Gamma[k,l] = \frac{1}{|\Gamma|} \sum_{[p,q]\in\Gamma} \delta[p,q], \quad \Gamma = \Gamma_t, \Gamma_f. \tag{45}$$

Now, the reliability index $\eta[k, l]$ is calculated by

$$\eta[k,l] = \exp\left\{-\min\left(\sigma_{\Gamma_t}[k,l], \sigma_{\Gamma_f}[k,l]\right)\right\} \tag{46}$$

where $\eta[k, l]$ is a normalized index satisfying $0 < \eta \leq 1$. When at least $\sigma_{\Gamma t}[k, l]$ or $\sigma_{\Gamma f}[k, l]$ at $[k, l]$ is sufficiently small, $\eta[k, l]$ approaches unity, thereby the corresponding delay value $\delta[k, l]$ is considered to be reliable. We observed the tendency that the PD error decreases as the reliability index increases. Then, the cell group is selected with reliability index $\eta[k, l] > \eta_{th}$ for subsequent DOA estimation. In this paper, η_{th} is set to 0.96. The reason for using this value and related remarks are given in later.

For each selected reliable T-F cell, the direction θ is computed using Eq.(41). Here the set of computed directions is denoted as follows:

$$\left\{\theta_i^{[l_i]} \middle| i = 1,2,\ldots,I\right\}, \tag{47}$$

where i is the numbering integer of the selected cells, I is the total number of data, and l_i is the frequency bin at which the i-th cell is located.

DOA error distribution model

Consider a T-F cell at which the n-th source dominates and is located in the unknown direction θ_n. From Eq. (41), the theoretical PD at the cell is given by

$$\phi_n[l] = \Delta\omega T l \sin\theta_n = B_n l, \tag{48}$$

where $B_n = \Delta\omega T \sin\theta_n$. The frame index k is omitted because k is not essential in this section. In the l-th frequency bin, the observed $\phi_n[l]$ is distributed around its mean value $B_n l$,

$$\phi_n[l] = B_n I + \Delta\phi[l], \tag{49}$$

where $\Delta\phi[l]$ is a random variable representing the PD estimation error. Then, assume that the random variable $\Delta\phi[l]$ is an independent identical Gaussian distribution with zero mean and constant variance σ_ϕ^2, that is, $N(o, \sigma_\phi^2)$. The constant variance means that $\Delta\phi[l]$ is independent of the frequency bin l; this assumption is represented as follows:

$$\Delta\phi[l] \sim N\left(0, \sigma^2{}_\phi\right). \tag{50}$$

Fig. 7 (a) illustrates Gaussian error distribution at l-th frequency bin in PD-F plane in two-source case. The Gaussian distribution assumption is motivated from the simplicity of theoretical manipulation. From these error distribution model the problem is to estimate the probability distribution of the direction θ as shown in Fig.7(b).

Now, the following proposition can be proved.

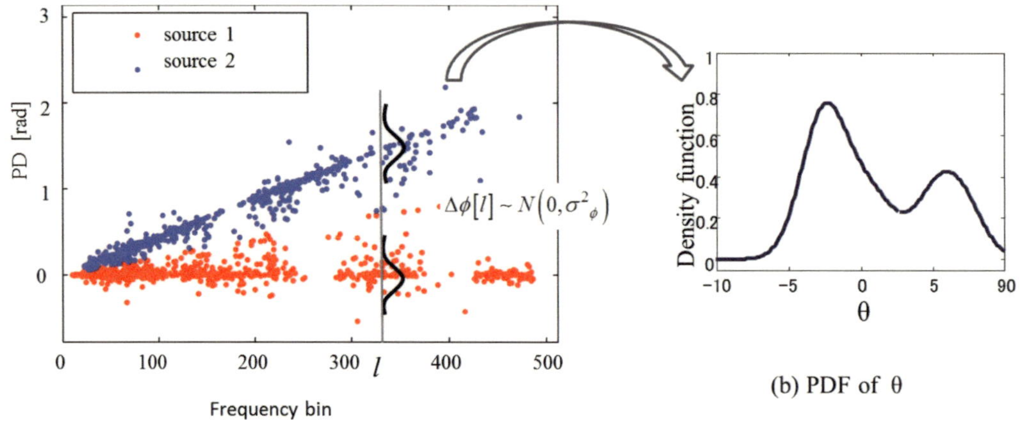

(a) Error distribution of PD

Figure 7. PD error distribution and Kernel density estimation

Proposition: If the random variable $\Delta\phi[l]$ is given by (50) and σ_ϕ is sufficiently small, the PDF of $\theta_n^{[l]}$ is given by

$$\theta_n^{[l]} \sim N\left(\theta_n, \sigma_{\theta_n}^2[l]\right), \tag{51}$$

$$\sigma_{\theta n}[l] = \frac{1}{T\Delta\omega l \cos\theta_n}\sigma_\phi. \tag{52}$$

This proposition can be proved by the linearized incremental analysis between $\phi[l]$ and $\theta^{[l]}$. The DOA error distribution model is shown in Fig. 8.

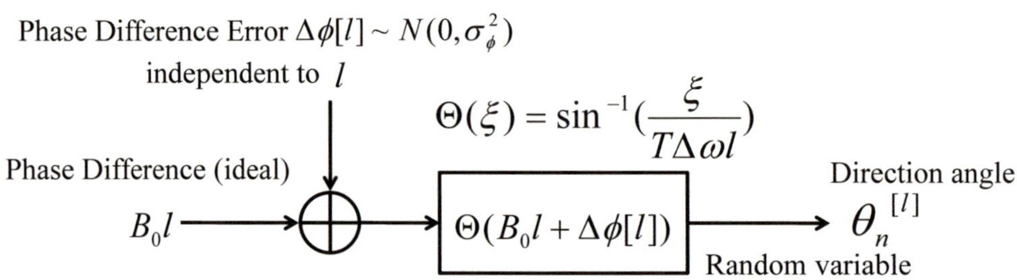

Figure 8. PD error and DOA estimation error distributions

4.2. DOA estimation using kernel density estimator

The kernel density estimation algorithm known as Parzen window in machine learning [32] is useful for statistical estimation even for a multiple-source problem. The algorithm provides an estimate PDF of $\theta[l]$ by using the observed samples (47). The maximum PDF point or the mode of the PDF can be considered as the optimal estimate of θ_n in the sense of the most probable value. The kernel density estimator approach yields an approximate estimation of the PDF of $\theta[l]$.

It is necessary to generalize the theoretical investigation noted above multisource and multi-frequency cases. The theoretical PDF formulation of θ in the case of multiple sources should be a Gaussian mixture with the same number of local modes (local peaks), each of which corresponds to an individual source. For the selected reliable data in Eq. (47), the kernel density estimator is applied to estimate the multi-modal PDF as follows:

$$\hat{p}(\theta) = \frac{1}{I}\sum_{i=1}^{I} \frac{1}{\varepsilon[l_i]} K\left(\frac{\theta - \theta_i^{[l_i]}}{\varepsilon[l_i]}\right), \tag{53}$$

where $K(\theta)$ is a kernel function, for which a Gaussian function is adopted in this study. $\varepsilon[l]$ is the bandwidth of the kernel. The idea behind applying the kernel density estimator is to reflect the theoretical result represented by the above proposition for the determination of the bandwidth. Since the variance of $\theta[l]$ depends on l and θ_n as indicated in Eq. (52), the bandwidth is determined as the form of

$$\varepsilon[I_i] = \frac{1}{T\Delta\omega l_i \cos\theta_i^{[I_i]}} \hbar. \tag{54}$$

where $\hbar$ is the control parameter and the observed $\theta[l_i]$ is substituted in place of a real unknown θ_n in Eq. (52). Accordingly, the dependence of the bandwidth on θ_n is indirectly controlled. The control parameter $\hbar$ is predetermined experimentally. Fig. 9 shows three examples of estimated PDFs for a two-source case with different $\hbar$. Finally, by finding the same number of local modes (peaks) as the number of pre-assigned source numbers, the source directions are estimated.

4.3. Experiments

Some experiments were conducted by the same setup and parameters as shown in Tab. 1. The first experiment is the case of two sources one of which is placed at the broad side (near 0 degree) as shown in Fig.10 (a). The results are shown in Fig.10 (b) and (c). While the proposed method gives a non-biased estimation, the estimates of the conventional method [27] tend to be biased for the cases of non-symmetric source positions with respect to the broadside. The second experiments for underdetermined case of three sources were performed. In this case three sources were symmetrically located at the closer locations { -23, 4, 23 degrees} and far apart locations{ -42, 4, 42 degrees}. Fig.11 (a) and (b) show the results of the conventional method [27] and the proposed. In the "far apart" case both methods can estimate the source directions well. However, for the "closer" case, the proposed method provides less biased estimates than [27]. From the additional experimental results with diffuse noise presented in [33] and [34] it is proved the proposed cell selection method provides noise robust estimation better than the conventional.

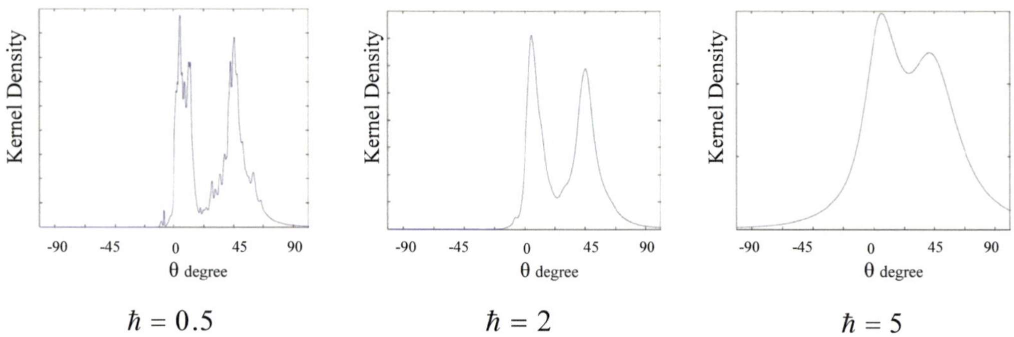

Figure 9. Estimated PDF and $\hbar$

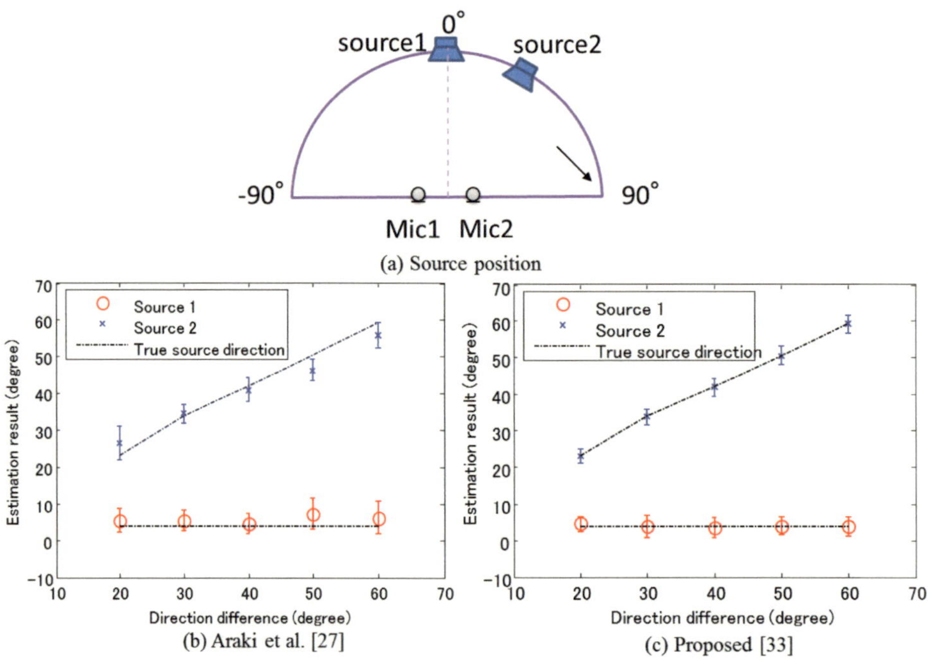

Figure 10. DOA estimation results for two sources

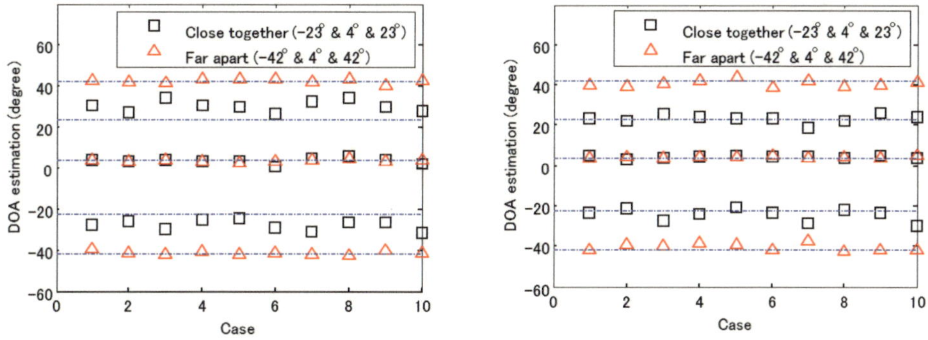

(a) Araki et al. [27] (b) Proposed [33]

Figure 11. DOA estimation results for three sources

5. Conclusions

This monograph summarizes speech segregation and speaker's direction estimation methods which are based on sparseness of T-F components of speech signals. Throughout the discussion we are interested in underdetermined source-sensor conditions. At first recent progresses on BSS and DOA estimation algorithms associated with T-F sparse representation are reviewed. Then we focus on presenting an author's solution of BSS problems exploiting a series of phase difference versus frequency data. In the algorithm time frame classification concerning source active states is performed, and actual separation procedure is solely applied to the mixing frames.

The latter half of this chapter treats DOA estimation algorithm in a pair of microphones.

The basic error propagating mechanism is introduced and then the kernel density estimator is applied. The method provides a robust and non-biased DOA estimation and it develops theory for arbitrary microphone array configuration. [35]

One of recent human machine speech communication research on segregation and localization is associated with robot auditory system where the tracking of moving sources and sensors have to be considered.[36] For coping with these cases the particle filter and adaptive array processing have been attractive, and further efforts will be made.

Acknowledgements

The authors would like to appreciate Professor Wlodzimierz Kasprzak of Warsaw University of Technology for his valuable suggestions, and all members of speech signal processing group of Hamada Laboratory in Keio University for their great help.

Author details

Nozomu Hamada and Ning Ding

Keio University, System Design Engineering, Faculty of Science and Technology, Japan

References

[1] Divenyi P., editor. Speech Separation by Humans and Machines. Kluwer Academic Publishers; 2005.

[2] Benesty J., Chen J., Huang Y. Microphone Array Signal Processing. Springer; 2008.

[3] Makino S, Lee TW, Sawada H., editors. Blind Speech Separation. Springer; 2007.

[4] Hyvarinen A., Karhumen J., Oja E. Independent Component Analysis. John Wiley & Sons, Inc.; 2001.

[5] Saruwatari S., Takatani T., Shikano K. SIMO-Model-Based Blind Source Separation - Principle and its Applications. In: Makino S et al. (ed.) Blind Speech Separation. Springer; 2007. p149-168.

[6] Sawada H., Araki S., Makino S. Frequency-domain Blind Source Separation. In: Makino S et al. (ed.) Blind Speech Separation. Springer; 2007. p47-78.

[7] Choi S., Lyu Y., Berthommier F., Glotin H., Cichoki A. Blind separation of delayed and superimposed acoustic sources: learning algorithms an experimental study. Proc. IEEE Int. Conference on Speech Processing (ICsP), Seoul 1999.

[8] Choi S., Hong H., Glotin H., Berthommier F. Multichannel signal separation for cocktail party speech recognition: A dynamic recurrent network. Neurocomputing 2002; 49 (1) 299-314.

[9] Huang J., Ohnishi N., Sugie N. A biomimetic system for localization and separation of multiple sound sources. IEEE Trans. on Instrumentation and Measurement 1995; 44(3) 733-738.

[10] Aoki M, Okamoto M, Aoki A, Matsui H, Sakurai T, Kaneda Y. Sound source segregation based on estimating incident angle of each frequency component of input signals acquired by multiple microphones. Acoust. Sci.& Tech 2001; 22(2) 149-157.

[11] Yilmaz O, Rickard S. Blind Separation of Speech Mixtures via Time- Frequency Masking. IEEE Trans. On signal processing 2004; 52(7) 1830-1847.

[12] Rickard S. The DUET Blind Source Separation Algorithm. In: Makino S et al. (ed.) Blind Speech Separation. Springer; 2007. p217-241.

[13] Araki S., Sawada H., Murai R., Makino S. Underdetermined blind sparse source separation for arbitrarily arranged multiple sensors. Signal Processing 2007; 87( ) 1833-1847.

[14] Sawada H., Araki S., Murai R., Makino S. Grouping separated frequency components by estimating propagation model parameters in frequency-domain blind source separation. IEEE Trans. on Audio, Speech, and Language Processing 2007 15(5) 1592-1604.

[15] Plumbley MD., Blumensath T., Daudet L., Gribonval R., Davies ME. Sparse representations in audio and music From coding to source separation. Proceedings of the IEEE 2010; 98(6) 995–1005.

[16] Nakatani T., Okuno H. Harmonic sound stream segregation using localization and its application to speech to speech stream segregation. Speech Communication 1999; 27 209-222.

[17] Parsons TW. Separation of speech from interfering speech by means of harmonic selection. Journal of the Acoustical Society of America 1976; 60(4) 911-918.

[18] Rickard S, Balan R., Rosca J. Real-time time frequency based blind source separation. ICA2001 2001; 651-656

[19] Izumi Y., Ono N., Sagayama S. Sparseness-based 2ch BSS using the EM algorithm in reverberant environment. Proc. IEEE Workshop Applications of Signal Processing to Audio and Acoustics 2007; 147–150.

[20] Benesty J., Chen J., Huang Y. Direction of Arrival and Time-Difference-of-Arrival Estimation. chapter 9 in Microphone Array Signal Processing, Springer, 2008.

[21] Claudio EDD., Parisi R. Multi-Source Localization Strategies. In: (ed.) Microphone Arrays. Springer-Verlag; 2001. p181–201.

[22] Knapp CH., Carter GC. The generalized correlation method for estimation of time delays. IEEE Trans. on Acoust. Speech Signal Process. 1976; ASSP24 320–327. .

[23] Schmidt RO. Multiple emitter location and signal parameter estimation. IEEE Trans. on Antennas and Propagation. 1986; 34 276–280.

[24] Abrard F., Deville Y. A time-frequency blind signal separation method applicable to underdetermined mixtures of dependent sources. Signal Processing 2005; 85 1389-1403.

[25] Arberet S., Gribonval R., Bimbot, "A robust method to count and locate audio sources in a multichannel underdetermined mixture," IEEE Trans. on Signal Processing, Vol. 58, No. 1, pp. 121-133, Jan. 2010.

[26] Berdugo B., Rosenhouse J., Azhari H. Speaker's direction finding using estimated time delays in the frequency domain. Signal Processing, 2002; 82 19–30.

[27] Araki S., Sawada H., Mukai R., Makino S. DOA estimation for multiple sparse sources with arbitrarily arranged multiple sensors. Journal of Signal Processing Systems, 2009; 63 265–275.

[28] Nesta F., Svaizer P., Omologo M. Cumulative state coherence transform for a robust two-channel multiple source localization. Proc. of ICA 2009; 290–297.

[29] Glotin H., Berthommier FB., Tessier E. A CASA-Labeling Model using the Localization Cue for Robust Cocktail-party Speech Recognition. Sixth European Conference on Speech Communication and Technology 1999; 22

[30] Tessier E., Berthommier F., Glotin H., Choi S. A CASA front-end using the localization cue for segregation and Then Cocktail-Party Speech Recognit6ion. Proc. IEEE Int. Conference on Speech Processing (ICsP) 1999; Seoul

[31] Ding N., Yoshida M., Ono J., Hamada N. Blind Source Separation Using Sequential Phase Difference versus Frequency Distortion. Journal of Signal Processing 2011; 15(5) 375-385.

[32] Duda R., Hart PE., Stork DG. Pattern Classification. John Wiley & Sons 2001.

[33] DING N., Hamada N. DOA Estimation of Multiple Speech Source from a Stereophonic Mixture in Underdetermined Case", IEICE Trans. Fundamentals, Vol.E95-A, No.4, Apr. 2012

[34] Ding N. Blind Source Separation and Direction Estimation for StereophonicMixtures of Multiple Speech Signals Based on Time-Frequency Sparseness. PhD thesis. Keio University Yokohama; 2012

[35] Fujimoto K., Ding N., Hamada N. Multiple Sources' Direction Finding by using Reliable Component on Phase Difference Manifold and Kernel Density Estimator. IEEE Proc. ICASSP 2012; Kyoto

[36] Valin JM., Michaud F., Rouat J. Robust localization and tracking of simultaneous moving sound sources using beamforming and particle filtering. Robotics and Autonomous System 2007; 55 216—228.

[37] Daobilige Su, Masashi Sekikawa, and Nozomu Hamada, Novel scheme of real-time direction finding and tracking of multiple speakers by robot-embedded microphone array, 1st Int. Con. on Robot Intelligence Tech. RiTA, 2012 Korea

Chapter 2

Evaluation of an Active Microphone with a Parabolic Reflection Board for Monaural Sound-Source-Direction Estimation

Tetsuya Takiguchi, Ryoichi Takashima and
Yasuo Ariki

Additional information is available at the end of the chapter

http://dx.doi.org/10.5772/56045

1. Introduction

For human-human or human-computer interaction, the talker's direction and location are important cues that determine who is talking. This information can be helpful, especially in multi-user conversation scenarios such as a meeting system, robotic communication, and so on. There have been studies for understanding of a conversation scene based on the talker localization approach (e.g., [1, 2]). An approach using the turn-taking information obtained from DOA (Direction-of-Arrival) estimation results for the discrimination of system requests or users' conversations has also been proposed [3].

Many systems using microphone arrays have been tried in order to localize sound sources. Conventional techniques, such as MUSIC (MUltiple SIgnal Classification), CSP (Cross-power Spectrum Phase), and so on (e.g., [4–9]), use simultaneous phase information from microphone arrays to estimate the direction of the arriving signal. There have also been studies on binaural source localization based on interaural differences, such as interaural level difference and interaural time difference (e.g., [10, 11]). Sound source localization techniques focusing on the auditory system have also been described in [12, 13].

Single-microphone source separation is one of the most challenging scenarios in the field of signal processing, and some techniques have been described (e.g., [14–17]). In our previous work [18], we discussed a sound source localization method using only a single microphone. In that report, the acoustic transfer function was estimated from observed (reverberant) speech using the statistics of clean speech signals without using texts of the user's utterance, where a GMM (Gaussian Mixture Model) was used to model the features of the clean speech. This estimation is performed in the cepstral domain employing a maximum-likelihood-based approach. This is possible because the cepstral parameters are an effective representation for retaining useful clean speech information. The experiment results of our talker-localization showed its effectiveness. However, the previous method required the measurement of speech for each room environment in advance. Therefore, this chapter presents a new method that uses parabolic reflection that is able to estimate the sound source direction without any need for such prior measurements.

In this chapter, we introduce the concept of an active microphone that achieves a good combination of active-operation and signal processing. The active microphone has a parabolic reflection board, which is extremely simple in construction. The reflector and its associated microphone rotate together, perform signal processing, and seek to locate the direction of the sound source. We call this microphone with the function of the rotation an active microphone.

A simple signal-power-based method using a parabolic antenna has been proposed in the radar field. But the signal-power-based method is not effective for finding the direction of a person talking in a room environment. One of the reasons is that the power of the speech signal varies for all directions of the parabolic antenna, since a person does not utter the same power (word) for all directions of the parabolic antenna. Therefore, in this chapter, our new sound-source-direction estimation method focuses on the acoustic transfer function instead of the signal power. The use of the parabolic reflection board results in a difference in the acoustic transfer functions of the target direction and the non-target directions, where the active microphone with the parabolic reflection board rotates and observes the speech at each angle. The sound source direction is detected by comparing the acoustic transfer functions observed at each angle, which are estimated from the observed speech using the statistics of clean speech signals. We compared our proposed method with the signal-power-based method, and as the methods for obtaining the directivity of the microphone, we compared the use of the parabolic reflection board with the use of a shotgun microphone. Its effectiveness is confirmed by sound-source-direction estimation experiments in a room environment.

2. Active microphone

2.1. Parabolic reflection board

In this chapter, an active microphone with a parabolic reflection board is introduced for estimation of sound source direction, where the reflection board has the shape of a parabolic surface. The parabolic reflector has been used for estimation of the direction of arrival in the radar field [19]. As shown in Figure 1, under the assumptions associated with plane waves, any line (wave) parallel to the axis of the parabolic surface is reflected toward the focal point. On the other hand, if the sound source is not located at 90 degrees (in front of the parabolic surface), no reflection wave will travel toward the focal point. Therefore, the use of the parabolic reflection board will be able to give us the difference in the acoustic transfer function between the target direction and the non-target directions.

2.2. Signal-power-based estimation of sound source direction

In [20], a simple signal-power-based method using a parabolic reflection board has been described. The use of parabolic reflection can increase the power gain of the signal arriving from directly in front of the parabolic board. In that method, the microphone with a parabolic

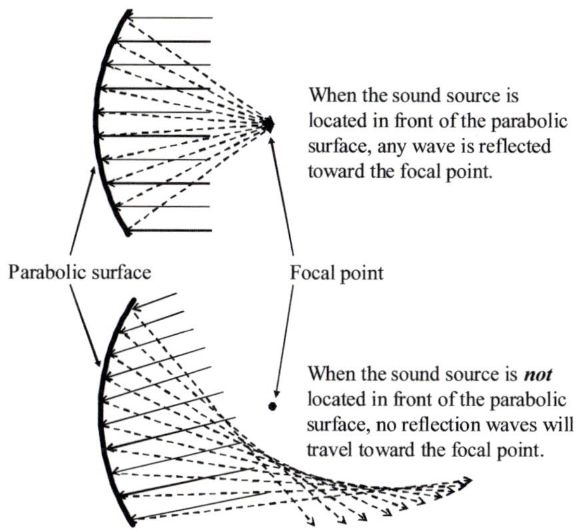

Figure 1. Concept of parabolic reflection

reflection board rotates, and calculates the power of the observed signal for each angle of the parabolic reflection board. Then, the direction having maximum power was selected as the sound source direction:

$$\hat{i} = \underset{i}{\text{argmax}} \sum_{n} \sum_{\omega} \log |O_i(\omega; n)|^2. \tag{1}$$

Here, $O(\omega; n)$ is the ω-th frequency bins of short-term linear spectrum at the frame n. i is the angle of the parabolic reflection board (microphone).

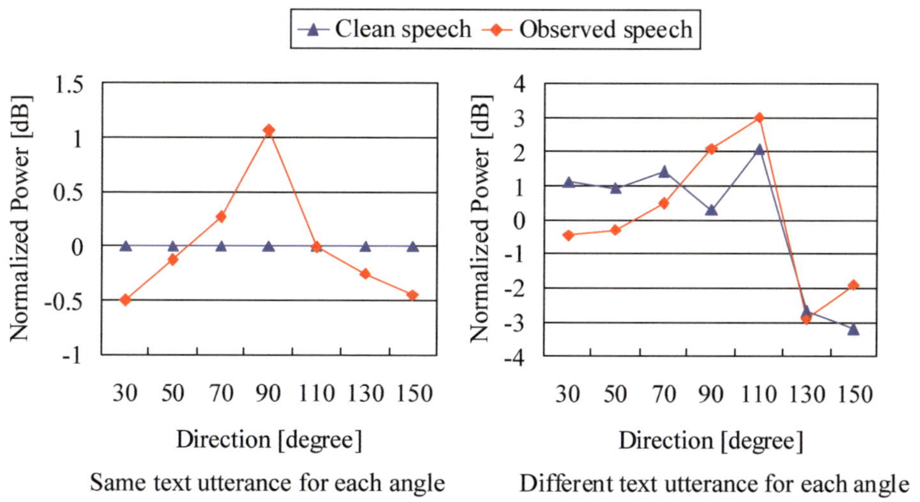

Figure 2. Power of a clean speech segment and the speech segment observed by the microphone with a parabolic reflection board for each angle. The power was normalized so that the mean values of all directions was 0 dB.

Its effectiveness has been confirmed on white noise signals. However, the signal-power-based method was not effective for finding the direction of a talking person. This is because the power of the uttered speech signals varies for all directions of the parabolic reflection board. Figure 2 shows the power of a clean speech segment and the observed speech segment for each angle of the parabolic reflection board. The size of the recording room was about 6.3 m × 7.2 m (width × depth). The target sound source was located at 90 degrees and 2 m from the microphone. The diameter of the parabolic reflection board was 24 cm, and the distance to the focal point was 9 cm. The speech signal was sampled at 12 kHz, and windowed with a 32-msec Hamming window every 8 msec. The power was normalized so that the mean values of all directions were 0 dB. In the left portion of Figure 2, the text utterance is the same for each angle of the parabolic reflection board, and in the right portion, the text utterance is different for each angle. As shown in this figure, the power of the observed speech was most enhanced by the parabolic reflection board at 90 degrees (target direction). However, when the utterance text differs at each angle of the parabolic reflection board, the power of observed speech at 90 degrees did not have the maximum power since the power of input speech for another direction had a higher power than that for 90 degrees. For this case, the signal-power-based fails to estimate the direction of the sound source correctly.

In this chapter, in order to estimate the direction of the sound source correctly when the power of the uttered speech signals varies for all direction of the parabolic reflection board, the acoustic transfer function is used instead of the power. Since the acoustic transfer function does not depend on the uttered clean speech, the use of the acoustic transfer function can estimate the direction of the sound source without the influence of the varying power of the uttered speech signals.

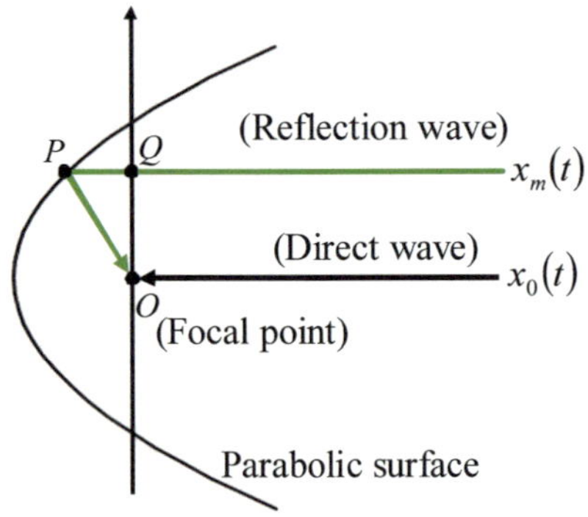

Figure 3. Observed signal at the focal point, where the input signal is coming from directly in front of the parabolic surface

2.3. Signal observed using parabolic reflection

Next, we consider the signal observed using parabolic reflection [20]. As shown in Figure 3, when the sound source is located directly in front of the parabolic surface and there are no background noise and no directivity of the sound source, the observed signal at the focal point at discrete time t can be expressed by the addition of the waves arriving at the focal point directly (direct wave) and those arriving at the focal point after being reflected by the parabolic surface (reflection waves):

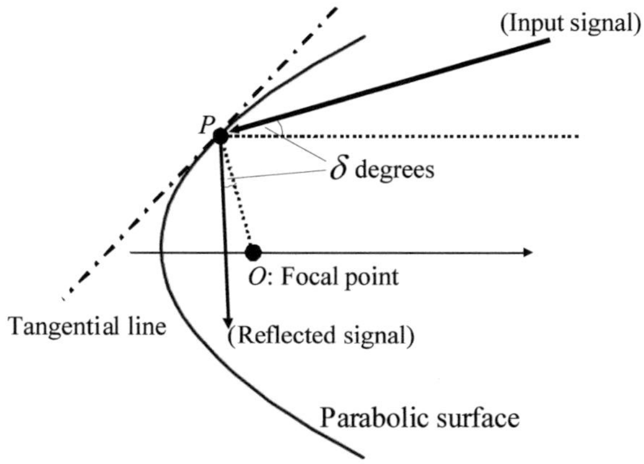

Figure 4. Observed signal at the focal point, where the input signal is coming from δ degrees

$$o(t) = x_p(t) + \sum_{m=1}^{M} x_m(t) \tag{2}$$

where $o(t)$, x_p and x_m ($m = 1, \cdots, M$) are observed sound, direct sound and reflection sound, respectively. Based on the properties of a parabola, the time difference to the focal point between the direct and reflection waves is constant without depending on m. Therefore, (2) can be written as

$$o(t) = s(t) * h_p(t) + \sum_{m=1}^{M} s(t-\tau) * h_m(t) \tag{3}$$

where $s(t)$ and τ are clean speech and the time difference, respectively. h_p is the acoustic transfer function of a direct wave and h_m is that of a reflection wave. By applying the short-term Fourier transform, the observed spectrum at frame n is given by

$$\begin{aligned} &O(\omega; n) \\ &\approx S(\omega; n) \cdot (H_p(\omega; n) + e^{-j2\pi\omega\tau} \cdot \sum_{m=1}^{M} H_m(\omega; n)) \\ &= S(\omega; n) \cdot (H_p(\omega; n) + H_r(\omega; n)). \end{aligned} \tag{4}$$

Here H_p is the acoustic transfer function of the direct sound that is not influenced by parabolic reflection. H_r is the acoustic transfer function resulting from parabolic reflection.

On the other hand, as shown in Figure 4, when the input signal is coming from δ degrees (not coming from directly in front of the parabolic surface), the direction of the reflected signal at the parabolic surface is off δ degrees from PO. Therefore, when the sound source is not located in front of the parabolic surface, parabolic reflection does not influence the acoustic transfer function since no reflection waves will travel toward the focal point:

$$O(\omega; n) \approx S(\omega; n) \cdot H_p(\omega; n). \tag{5}$$

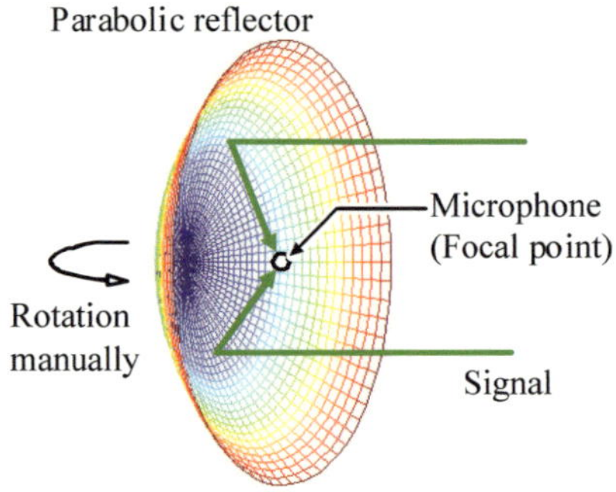

Figure 5. Active microphone with parabolic reflection

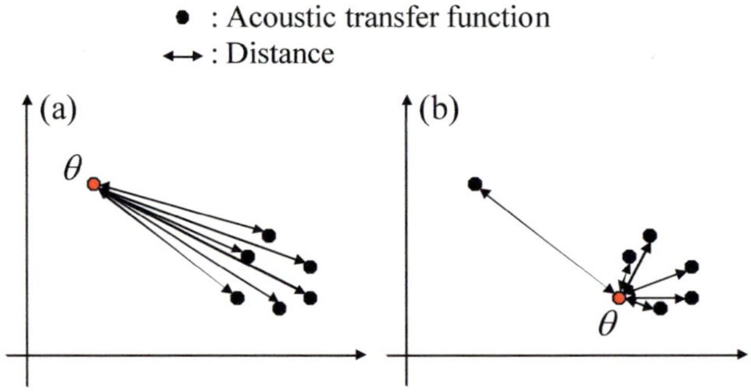

Figure 6. Acoustic transfer function in a feature space for each angle of the active microphone. (a) The case that the direction θ has the acoustic transfer function which is the farthest from those of other directions. (b) The case that the acoustic transfer function of θ is similar to most of the acoustic transfer functions of other directions.

2.4. Estimation of sound source direction

As shown in Figure 5, new active microphone with a parabolic reflection board was constructed with the microphone located at the focal point. In order to obtain the signal observed at each angle, the angle of the microphone was changed manually in research carried out for this chapter. Then, from equations (4) and (5), the spectrum of the signal observed at a microphone angle θ can be expressed as

$$
\begin{aligned}
O_\theta(\omega;n) &\approx S_\theta(\omega;n) \cdot H_\theta(\omega;n) \\
H_\theta(\omega;n) &= \begin{cases} H_p(\omega;n) + H_r(\omega;n) & (\theta = \hat{\theta}) \\ H_p(\omega;n) & (\theta \neq \hat{\theta}) \end{cases}
\end{aligned} \tag{6}
$$

where S_θ and H_θ are spectra of clean speech and acoustic transfer function at the angle θ, and $\hat{\theta}$ is the sound source direction. Assuming H_p is nearly constant for each angle, when the active microphone does not face the sound source, the value of H_θ will be almost the same for every non-target direction. On the other hand, the only condition under which H_θ will have a different value from that obtained at the other angles is when the active microphone faces the sound source.

Therefore, the acoustic transfer function is estimated at each discrete direction θ, and the sound source direction can be estimated by selecting the direction whose the acoustic transfer function is the farthest from the acoustic transfer functions of other directions. The sum of the mutual distances is used to find such a direction. For each discrete direction θ, the Euclidean distances from the acoustic transfer function of θ to those of other directions are measured. As shown in Figure 6, when the direction θ has the acoustic transfer function which is the farthest from those of other directions, the sum of the distances becomes larger than those of other directions. Hence, the sound source direction is estimated by selecting the direction having the maximum sum of the distances from the acoustic transfer function of the direction to those of other directions:

$$\hat{\theta} = \underset{\theta}{\text{argmax}} \sum_{\theta'} (\bar{H}_{\theta} - \bar{H}_{\theta'})^2 \tag{7}$$

where θ' is all directions of the microphone except θ, and $\bar{H}$ is the expectation of H in regard to the time frame. Actually, in this research, the cepstrum of acoustic transfer function is used to calculate this equation. In the next section, we will describe how to estimate H_{θ} from observed speech signals.

3. Estimation of the acoustic transfer function

In our previous work [18], we proposed a method to estimate the acoustic transfer function from the reverberant speech (any utterance) using the clean-speech acoustic model, where a GMM is used to model the feature of the clean speech. The clean speech GMM enables us to estimate the acoustic transfer function from the observed speech without needing to have texts of the user's utterance (text-independent estimation). However, because an active microphone with parabolic reflection board was not used, the previous method required the measurement of speech for each room environment in advance in order to be able to determine the direction of a talking person. In this chapter, we can estimate the sound source direction without any need for prior measurements by information fusion of an active microphone and an estimation of an acoustic transfer function.

3.1. Cepstrum representation of reverberant speech

The reverberant speech signal, $o(t)$, in a room environment is generally considered to be the convolution of clean speech and the acoustic transfer function $o(t) = \sum_{l=0}^{L-1} s(t-l)h(l)$, where $s(t)$, $h(l)$ and L are a clean speech signal, an acoustic transfer function (room impulse response) from the sound source to the microphone, and the length of the acoustic transfer function, respectively.

In recent studies for robust speech recognition and speech dereverberation, the reverberant speech in the STFT (Short-Term Fourier Transform) domain is often modeled so that each frequency bin of the reverberant speech is represented by the convolution of the frame sequences of clean speech and the acoustic transfer function [21, 22].

$$O(\omega; n) = \sum_{l'=0}^{L'-1} S(\omega; n-l') \cdot H(\omega; l') \tag{8}$$

Here $O(\omega;n)$, $S(\omega;n)$, and $H(\omega;n)$ are the ω-th frequency bins of short-term linear spectra of the frame n. L' is the length of the acoustic transfer function in the STFT domain. However, that modeling is complex for estimating the frame sequence of the acoustic transfer function, and it is difficult to deal with the estimated components of the acoustic transfer function for this talker localization task. Therefore, in this chapter, we employ a simpler modeling of the reverberant speech, which is approximately represented as the product of clean speech and the acoustic transfer function.

$$O(\omega;n) \approx S(\omega;n) \cdot H(\omega;n) \tag{9}$$

Cepstral parameters are an effective representation for retaining useful speech information in speech recognition. Therefore, we use the cepstrum for acoustic modeling necessary to estimate the acoustic transfer function. The cepstrum of the observed signal is given by the inverse Fourier transform of the log spectrum:

$$O_{cep}(d;n) \approx S_{cep}(d;n) + H_{cep}(d;n) \tag{10}$$

where d is the cepstral index. O_{cep}, S_{cep}, and H_{cep} are cepstra for the observed signal, clean speech signal, and acoustic transfer function, respectively. As shown in equation (10), if O and S are observed, H can be obtained by

$$H_{cep}(d;n) \approx O_{cep}(d;n) - S_{cep}(d;n). \tag{11}$$

However S cannot be observed actually. Therefore H is estimated by maximizing the likelihood (ML) of observed speech using clean-speech GMM.

3.2. Maximum-likelihood-based parameter estimation

The sequence of the acoustic transfer function in (11) is estimated in an ML manner [23] by using the expectation maximization (EM) algorithm, which maximizes the likelihood of the observed speech:

$$\hat{H} = \underset{H}{\operatorname{argmax}} \Pr(O|H, \lambda_S). \tag{12}$$

Here, λ denotes the set of GMM parameters of the clean speech, while the suffix S represents the clean speech in the cepstral domain. The GMM of clean speech consists of a mixture of Gaussian distributions.

$$\lambda_S = \{w_k, N(\mu_k^{(S)}, \sigma_k^{(S)^2})\}, \ \sum_k w_k = 1 \tag{13}$$

where w_k, μ_k and σ_k^2 are the weight coefficient, mean vector and variance vector (diagonal covariance matrix) of the k-th mixture component, respectively. These parameters are estimated using the EM algorithm using a clean speech database.

The estimation of the acoustic transfer function in each frame is performed in a maximum likelihood fashion by using the EM algorithm. The EM algorithm is a two-step iterative procedure. In the first step, called the expectation step, the following auxiliary function Q is computed.

$$\begin{aligned} Q(\hat{H}|H) &= E[\log \Pr(O, c|\hat{H}, \lambda_S)|H, \lambda_S] \\ &= \sum_c \frac{\Pr(O, c|H, \lambda_S)}{\Pr(O|H, \lambda_S)} \cdot \log \Pr(O, c|\hat{H}, \lambda_S) \end{aligned} \tag{14}$$

Here c represents the unobserved mixture component labels corresponding to the observation sequence O.

The joint probability of observing sequences O and c can be calculated as

$$\Pr(O, c|\hat{H}, \lambda_S) = \prod_n w_{c(n)} \Pr(O(n)|c(n), \hat{H}, \lambda_S) \tag{15}$$

where w is the mixture weight and O_n is the cepstrum at the n-th frame. Since we consider the acoustic transfer function as additive noise in the cepstral domain, the mean to mixture k in the model λ_O is derived by adding the acoustic transfer function. Therefore, equation (15) can be written as

$$\Pr(O, c|\hat{H}, \lambda_S) = \prod_n w_{c(n)} \cdot N(O(n); \mu_k^{(S)} + \hat{H}(n),\ \Sigma_k^{(S)}) \tag{16}$$

where $N(O; \mu, \Sigma)$ denotes the multivariate Gaussian distribution. It is straightforward to derive that

$$\begin{aligned} Q(\hat{H}|H) &= \sum_k \sum_n \Pr(O(n), c(n) = k|H, \lambda_S) \log w_k \\ &\quad + \sum_k \sum_n \Pr(O(n), c(n) = k|H, \lambda_S) \\ &\quad \cdot \log N(O(n); \mu_k^{(S)} + \hat{H}(n), \Sigma_k^{(S)}). \end{aligned} \tag{17}$$

Here $\mu_k^{(S)}$ and $\Sigma_k^{(S)}$ are the k-th mean vector and the (diagonal) covariance matrix in the clean speech GMM, respectively. It is possible to train those parameters by using a clean speech database. Next, we focus only on the term involving H.

$$\begin{aligned} Q(\hat{H}|H) = -\sum_k \sum_n \gamma_k(n) \sum_{d=1}^{D} \Bigg\{ &\frac{1}{2} \log(2\pi)^D \sigma_{k,d}^{(S)^2} \\ &+ \frac{(O(d;n) - \mu_{k,d}^{(S)} - \hat{H}(d;n))^2}{2\sigma_{k,d}^{(S)^2}} \Bigg\} \end{aligned} \tag{18}$$

$$\gamma_k(n) = \Pr(O(n), k|H, \lambda_S) \tag{19}$$

Here $O(n)$ is the cepstrum at the n-th frame for observed speech data. D is the dimension of the $O(n)$, and $\mu_{k,d}^{(S)}$ and $\sigma_{k,d}^{(S)^2}$ are the d-th mean value and the d-th diagonal variance value of the k-th component in the clean speech GMM, respectively.

The maximization step (M-step) in the EM algorithm becomes "max $Q(\hat{H}|H)$". The re-estimation formula can, therefore, be derived, knowing that $\partial Q(\hat{H}|H)/\partial \hat{H} = 0$ as

$$\hat{H}(d;n) = \frac{\sum_k \gamma_k(n) \frac{O(d;n) - \mu_{k,d}^{(S)}}{\sigma_{k,d}^{(S)2}}}{\sum_k \frac{\gamma_k(n)}{\sigma_{k,d}^{(S)2}}}. \tag{20}$$

Therefore, the frame sequence of the acoustic transfer function $\hat{H}_\theta(d;n)$ is estimated from the signal $O_\theta(d;n)$ observed at direction θ in the cepstral domain using equation (20).

We obtain $\hat{H}_\theta(d;n)$ at a discrete direction. Next, the d-th dimension of the mean vector $\bar{H}_\theta(d)$ is obtained by averaging $\hat{H}_\theta(d;n)$ per frame n.

$$\bar{H}_\theta(d) = \sum_n \hat{H}_\theta(d;n) \tag{21}$$

In a similar way, we obtain the mean vector $\bar{H}_\theta(d)$ at all discrete directions, and the sound source direction is estimated using equation (7) using the cepstral vector $\bar{H}_\theta$. In this chapter, the angle of the parabolic reflection microphone was changed manually from 30 degrees to 150 degrees in increments of 20 degrees.

4. Experiment

4.1. Experiment conditions

The direction estimation experiment was carried out in a real room environment. The parabolic reflection microphone shown in Figure 5 was used for the experiments. The diameter was 24 cm, and the distance to the focal point was 9 cm. The microphone located at the focal point was an omnidirectional type (SONY ECM-77B). The target sound source was located at 90 degrees and 2 m from the microphone. The angle of the parabolic reflection microphone was changed manually from 30 degrees to 150 degrees in increments of 20 degrees. Then the acoustic transfer function of the target signal at each angle was estimated for the following speech lengths: 0.5, 1.0, 1.5, 2.0, 2.5, and 3.0 seconds. The size of the recording room was about 6.3 m $\times$ 7.2 m (width $\times$ depth). Figure 7 shows the environment of the experiment.

The speech signal was sampled at 12 kHz, and windowed with a 32-msec Hamming window every 8 msec. The clean speech GMM was trained by using 50 sentences (spoken by a female) in the ASJ Japanese speech database. The trained GMM has 64 Gaussian mixture components. For estimation of the acoustic transfer function from the observed speech

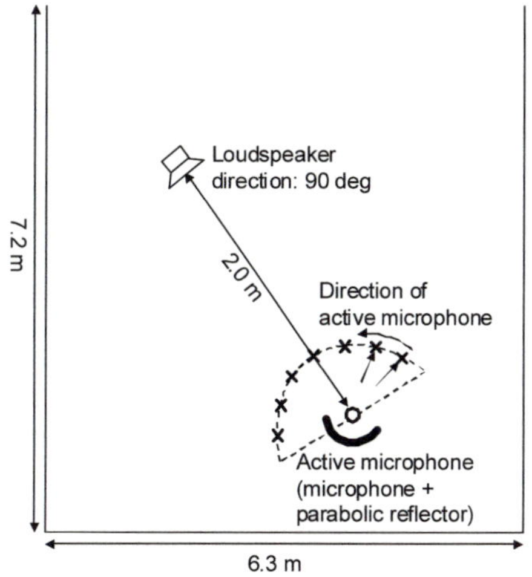

Figure 7. Experimental conditions

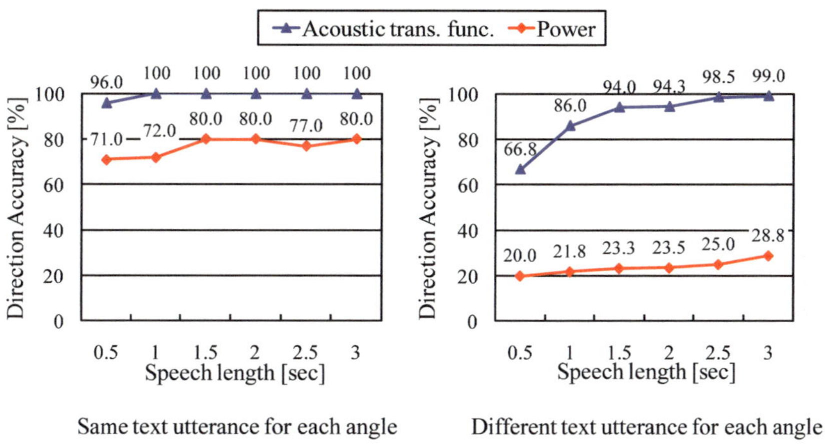

Figure 8. Performance of an active microphone with a parabolic reflection board

signal, 16-order MFCCs (Mel-Frequency Cepstral Coefficients) were used as feature vectors of the clean speech and estimated acoustic transfer function. Then, the 1st and 2nd orders of MFCCs of the estimated acoustic transfer function were used for estimating the sound source direction using equation (7). The test data was spoken by the same female who recorded the training data. The text utterances, however, were different.

4.2. Experiment results

Figure 8 shows the direction accuracy performance using the acoustic transfer function estimated at various speech lengths. The performance is compared to the power-based technique. The left figure shows the accuracy for the same text utterance at each angle of the active microphone, and the right figure shows the accuracy for a different text utterance at each angle of the active microphone. The test data for the same text utterance consisted of 100 segments. The test data for the different utterance consisted of 600, 300, 200, 150,

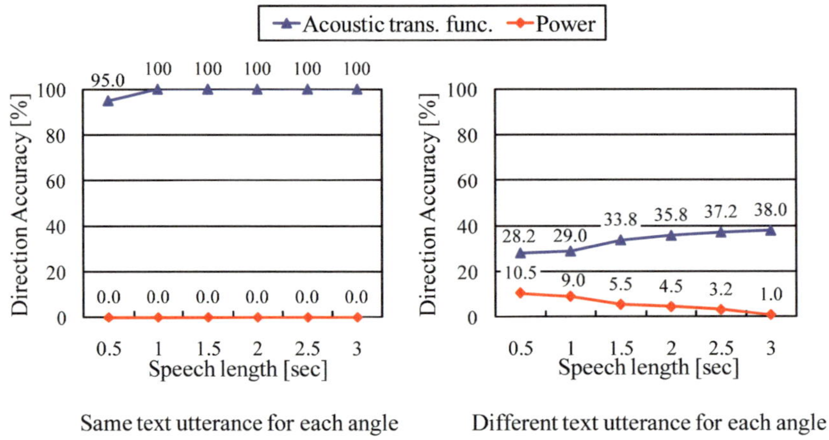

Figure 9. Performance of a shotgun microphone without a parabolic reflection board

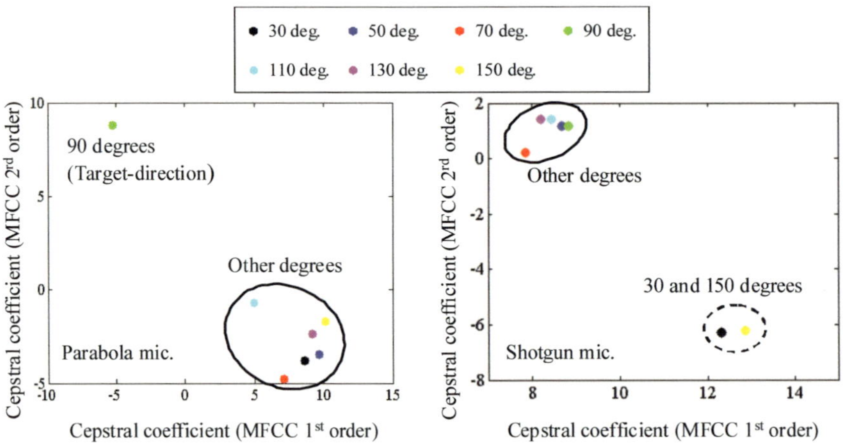

Figure 10. Mean values of the acoustic transfer functions for the microphone with a parabolic reflection board (left) and the shotgun microphone (right)

120, and 100 segments, where one segment has a time length of 0.5, 1.0, 1.5, 2.0, 2.5, and 3.0 seconds, respectively. The test for the same text utterance was conducted 100 times, and that for the different text utterance was conducted 600 times by changing the combination of the text utterances for each direction.

As shown in the left figure, the performance for both the techniques based on the power and the acoustic transfer function is high. However, the possibility of there being an identical text utterance at each angle of the active microphone will be very small in a real environment. In the right portion of Figure 8, we can see that the performance of the power-based technique degrades drastically when the utterance text differs at each angle of the active microphone, because the power of the speech signal varies for all directions of the active microphone.

On the other hand, the performance of the new method based on the acoustic transfer function is high, even for different text utterances. This is because the new method uses the information of the acoustic transfer function, which depends on the direction of the

http://dx.doi.org/10.5772/56045

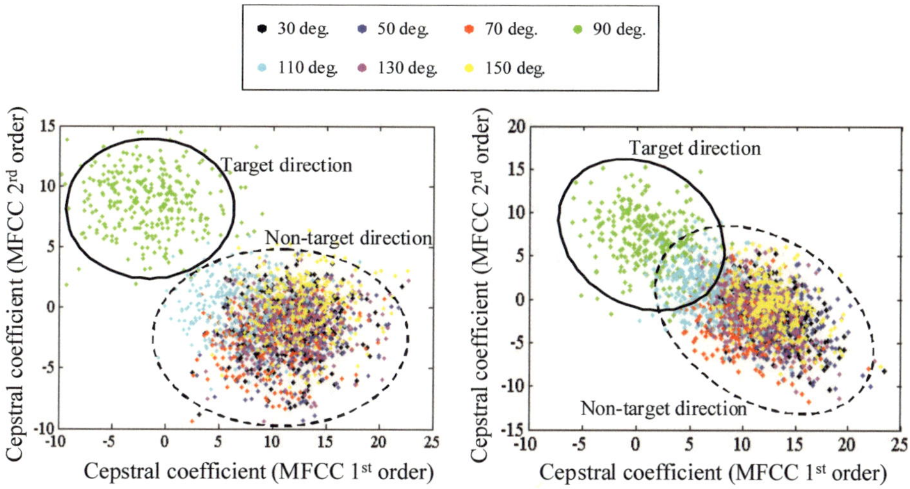

Acoustic transfer function computed by using true clean speech data (H_{sub})

Acoustic transfer function estimated by using the clean speech GMM (H_{est})

Figure 11. Acoustic transfer function computed by using true clean speech data (left) and that estimated by the proposed method using only the statistics of clean speech GMM (right) at each angle in the cepstral domain

active microphone only and does not depend on the utterance text. Also, we can see that the shorter the speech length for each angle is, the more the direction accuracy decreases. One reason for this is that the statistics for the observed speech are not readily available if there are not enough samples are used to estimate the acoustic transfer function.

Figure 9 shows the performance of a shotgun microphone (SONY ECM-674) without a parabolic reflection board. The power-based method can provide good performance for the same text utterance at each angle of the shotgun microphone due to the directivity of the shotgun microphone, but the performance degrades when the utterance text differs at each angle of the shotgun microphone. On the other hand, the performance of the new method based on the acoustic transfer function is even lower. The directivity of the shotgun microphone changes drastically as the sound-source direction changes from the front direction to the side directions of the shotgun microphone, and, as a result, the acoustic transfer function that is farthest from all the other acoustic transfer functions comes to be that at 30 or 150 degrees in equation (7). The mean values of all acoustic transfer functions for a parabolic reflection board and the shotgun microphone are plotted in Figure 10, where the acoustic transfer function is computed by (11) using true clean speech signal $S_{cep}(d;n)$ and the total number of frames is 36,600. Then the mean values are computed. As shown in the right potion of Figure 10, we can see that the acoustic transfer function that is farthest from all the other acoustic transfer functions is that at 30 or 150 degrees. As shown in the left potion of Figure 10, on the other hand, the acoustic transfer function that is farthest from all the other acoustic transfer functions is that at 90 degrees to the target direction.

Figure 11 shows the plot of acoustic transfer function for 300 segments of observed speech for the case of the active microphone. In the left potion of Figure 11, the acoustic transfer function H_{sub} was computed by (11) using true clean speech signal $S_{cep}(d;n)$. On the other hand, in the right potion of Figure 11, the acoustic transfer function H_{est} was estimated by

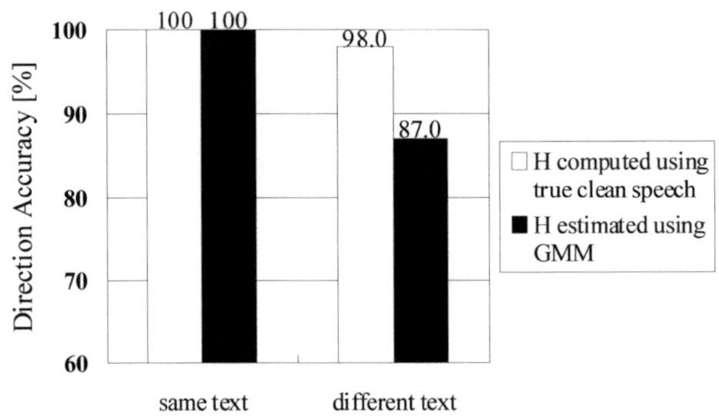

Figure 12. Comparison of true clean speech data and clean speech model

(20) using only the statistics of clean speech GMM. As shown in the left potion of Figure 11, when the active microphone does not face the sound source, H_{sub} is distributed in almost the same place, and H_{sub} of the sound source direction is distributed away from the H_{sub} of other directions. In the right potion of Figure 11, though the distribution of the estimated H_{est} may have some slight variations, it can be said that the distribution of H_{est} is similar that of H_{sub}.

Figure 12 shows the difference in the direction accuracy between the use of H_{sub} (the true clean speech data) and H_{est} (the statistics of clean speech model: GMM). As shown in this figure, when the utterances for each angle consist of the same text, the direction accuracy was 100%. However, when the texts of utterances for each angle are different, the direction accuracy obtained using H_{est} decreased. This is because the value of H_{est} was influenced to some extent by the phoneme sequence of clean speech.

5. Conclusions

This chapter has introduced the concept of an active microphone that achieves a good combination of active-operation and signal processing, and described a sound-source-direction estimation method using a single microphone with a parabolic reflection board. The experiment results in a room environment confirmed that the acoustic transfer function influenced by parabolic reflection can clarify the difference between the target direction and the non-target direction. In future work, more research will be needed in regard to different utterances and direction estimation in short intervals.

It is difficult for this method to estimate the directions of multiple sound sources because it is difficult to estimate the acoustic transfer functions of multiple sound sources. Also, the background noise and the reverberation may cause the measurement error of the acoustic transfer function. We will evaluate the performance of the proposed system in noisy environments and various room environments. In addition, we intend to investigate the performance of the proposed system when the directivity and the orientation of the sound source changes, and to test the performance of the system in a speaker-independent speech model.

Author details

Ryoichi Takashima, Tetsuya Takiguchi and Yasuo Ariki

Graduate School of System Informatics, Kobe University, Kobe, Japan

References

[1] X. Anguera, C. Wooters, and J. Hernando. Acoustic beamforming for speaker diarization of meetings. IEEE Trans. Audio, Speech and Language Processing 2007; 15 2011–2022.

[2] S. Araki, M. Fujimoto, K. Ishizuka, H. Sawada, and S. Makino. A DOA based speaker diarization system for real meetings. proceedings of the Hands-free Speech Communication and Microphone Arrays, HSCMA 2008, 29–32 May 2008, Trento, Italy.

[3] T. Yamagata, A. Sako, T. Takiguchi, and Y. Ariki. System request detection in conversation based on acoustic and speaker alternation features. proceedings of the Interspeech 2007, 2789–2792 August 2007, Antwerp, Belgium.

[4] D. Johnson and D. Dudgeon. Array Signal Processing. New Jersey: Prentice Hall; 1996.

[5] C. Knapp and G. Carter. The generalized correlation method for estimation of time delay. IEEE Trans. on Acoustics, Speech and Signal Processing 1976; 24 320–327.

[6] M. Omologo and P. Svaizer. Acoustic event localization in noisy and reverberant environment using csp analysis. proceedings of the International Conference on Acoustics, Speech, and Signal Processing, ICASSP 1996, 921–924 May 1996, Atlanta, Georgia.

[7] F. Asano, H. Asoh, and T. Matsui. Sound source localization and separation in near field. IEICE Trans. on Fundamentals of Electronics, Communications and Computer Sciences 2000; E83-A 2286–2294.

[8] Y. Denda, T. Nishiura, and Y. Yamashita. Robust talker direction estimation based on weighted CSP analysis and maximum likelihood estimation. IEICE Trans. on Information and Systems 2000; E89-D 1050–1057.

[9] J. H. DiBiase, H. F. Silverman, and M. S. Brandstein. Robust localization in reverberant rooms. In: Microphone Arrays: Signal Processing Techniques and Applications. Berlin, Grmany: Springer; 2001. p157–180.

[10] F. Keyrouz, Y. Naous, and K. Diepold. A new method for binaural 3-D localization based on HRTFs. proceedings of the International Conference on Acoustics, Speech, and Signal Processing, ICASSP 2006, vol.5 341–344 May 2006, Toulouse, France.

[11] M. Takimoto, T. Nishino, and K. Takeda. Estimation of a talker and listener's positions in a car using binaural signals. proceedings of the 4th Joint Meeting ASA and ASJ, ASA/ASJ06, 3216 November 2006, Honolulu, Hawaii.

[12] O. Ichikawa, T. Takiguchi, and M. Nishimura. Sound source localization using a pinna-based profile fitting method. proceedings of the International Workshop on

Acoustic Echo and Noise Control, IWAENC 2003, 263–266 September 2003, Kyoto, Japan.

[13] N. Ono, Y. Zaitsu, T. Nomiyama, A. Kimachi, and S. Ando. Biomimicry sound source localization with fishbone. IEEJ Trans. Sensors and Micromachines 2001; 121-E(6) 313–319.

[14] T. Kristjansson, H. Attias, and J. Hershey. Single microphone source separation using high resolution signal reconstruction. proceedings of the International Conference on Acoustics, Speech, and Signal Processing, ICASSP 2004, 817–820 May 2004, Quebec, Canada.

[15] B. Raj, M. V. S. Shashanka, and P. Smaragdis. Latent dirichlet decomposition for single channel speaker separation. proceedings of the International Conference on Acoustics, Speech, and Signal Processing, ICASSP 2006, vol.5 May 2006, Toulouse, France.

[16] G.-J. Jang, T.-W. Lee, and Y.-H. Oh. A subspace approach to single channel signal separation using maximum likelihood weighting filters. proceedings of the International Conference on Acoustics, Speech, and Signal Processing, ICASSP 2003, 45–48 Apr 2003, Hong Kong, China.

[17] T. Nakatani and B.-H. Juang. Speech dereverberation based on probabilistic models of source and room acoustics. proceedings of the International Conference on Acoustics, Speech, and Signal Processing, ICASSP 2006, I–821–I–824 May 2006, Toulouse, France.

[18] R. Takashima, T. Takiguchi, and Y. Ariki. Single-channel talker localization based on discrimination of acoustic transfer functions. In: P. Strumillo (ed.) Advances in Sound Localization. Rijeka: InTech; 2011. p39–54.

[19] B. Saka and A. Kaderli. Direction of arrival estimation and adaptive nulling in array-fed reflectors. proceedings of the Electrotechnical Conference, 274–277 1998.

[20] T. Takiguchi, R. Takashima, and Y. Ariki. Active microphone with parabolic reflection board for estimation of sound source direction. proceedings of the Hands-free Speech Communication and Microphone Arrays, HSCMA 2008, 65–68 May 2008, Trento, Italy.

[21] A. Sehr, R. Maas, and W. Kellermann. Reverberation model-based decoding in the logmelspec domain for robust distant-talking speech recognition. IEEE Trans. on Audio Speech and Language Processing 2010; 18 1676–1691.

[22] T. Nakatani, T. Yoshioka, K. Kinoshita, M. Miyoshi, and B.-H. Juang. Blind speech dereverberation with multi-channel linear prediction based on short time Fourier transform representation. proceedings of the International Conference on Acoustics, Speech, and Signal Processing, ICASSP 2008, 85–88 March 2008, Las Vegas, Nevada.

[23] B.-H. Juang. Maximum likelihood estimation for mixture multivariate stochastic observations of Markov chains. AT&T Technical Journal 1985; 64 1235–1250.

Chapter 3

Application of Iterative Reverse Time Migration Procedure on Transcranial Thermoacoustic Tomography Imaging

Zijian Liu and Lanbo Liu

Additional information is available at the end of the chapter

http://dx.doi.org/10.5772/56619

1. Introduction

Thermoacoustic tomography (TAT) [1]-[3] is a novel, noninvasive medical imaging technique that detects the large differences in microwave absorption between pathological and normal tissue [4], [5].

It applies the principle of the thermoacoustic effect [6], an effect that arises from the combination of the pressure oscillations of a sound wave with the accompanying adiabatic temperature oscillations [7]. Within TAT an input microwave impulse stimulates thermo-expansion in bio-tissue and consequently generates acoustic waves (P wave) to be recorded by transducers arranged outside the tissue. When the tissue is relatively uniform, the initial local acoustic amplitude is approximately proportional to the absorption ratio of the microwave [1], [8], [9]. Consequently, the TAT imaging problem is to retrieve the distribution of the initial acoustic amplitude based on recorded acoustic wave energy, and thus it can be characterized as a problem of "multiple sources localization".

In recent years, imaging by using thermoacoustic effect has raise increasing concerns [10], [11]. In which TAT has also been well studied for brain imaging [8], [9], [12]. The reasons include the following: first and far more important, TAT takes advantage of deep penetration of the electromagnetic impulse and the high resolution of the ultrasonic wave for deep imaging. Second, brain tissue is fundamentally uniform and isotropic. For example, the acoustic velocity of human brain is narrowly ranged within 1483-1521 m/s [13]. Acoustic wave propagation inside the brain is very close to one-way line-of-sight transmission without much aberration. The most commonly used imaging algorithm of TAT is to back-project the wave energy recorded by each transducer along the ray paths to all possible locations inside the imaging domain [1], [9]. The back-projection method [15]-[18] has been reported to be well performed on its simple scheme with high cost-effectiveness [19]. However, the approximation of one-way sight transmission used in the back-projection algorithm is no longer valid when high velocity contrast exists, for example, when the skull with an acoustic speed of 2500-2900 m/s [20] is included in the imaging domain. For compensating skull-related aberration Jin et al. [13] have developed a strategy based on the approximation of ray-tracing; however, it may still suffer from the difficulty of accurately calculating Green's function when the velocity structure becomes more irregular.

Currently reverse-time migration (RTM) has emerged as a more precise and powerful imaging tool in the exploration geophysics community [21]-[24]. RTM takes full advantage of the wave equation that includes all the dynamic features of a propagating wave field. Different from back-projection, RTM is based on the insensitivity of the wave equation's solution to the directionality of time. During RTM, by solving the wave equation with either the finite-differences time domain (FDTD) method [25] or the pseudo-spectral time domain method [26], all transducers act as a virtual source by broadcasting their own records back to the domain in a time-reversed manner. If the velocity model is precise, the reversed-time wave field should converge and be enhanced at the origins of the to-be-imaged structures. Previous studies [19], [27], [28] have compared back-projection method and RTM by using complex velocity structure models, and have confirmed better results by using RTM. However, with the advantage of RTM outlined previously, the barriers to RTM adoption are also evident. Compared with back-projection, RTM appears to be more sensitive to accuracy of given velocity model. Once the velocity model is inaccurate, the quality of imaging results from RTM may be insignificant in comparison to back-projection [19], [27]. Consequently the solutions for creating accurate velocity model become extremely vital to guarantee the imaging quality of RTM.

For estimate velocity model correctly before RTM, to the best of author's horizon there are two types of approaches. One is tomography, such as ultrasonic transmission tomography. Although it has been reported to be effectiveness for velocity correction from Jin and Wang in 2006 [29], prior of RTM detection, additional laboratory experiment has to be conducted for tomography data collection. This setup obviously increases the complexity of imaging procedure and financial cost. The other approach is an iterative imaging procedure developed by Whitmore [23]. In his approach for estimate the velocity correctly, an initial guess of the velocity model is setup by incorporating all known external velocity information as the first step. After that the TAT image result is derived by RTM, and a comparison between the reconstructed image and the velocity model is made. The differences derived by this comparison are attempted to be eliminated by generating the new velocity model. This procedure is repeated until both RTM results and velocity models are unchangeable. This iterative RTM procedure will be applied in this work.

Based on our previous studies [28], [30], [31] in this paper we closely concentrate on two topics: 1) comparison between back-projection and RTM on transcranial TAT; 2) Application of iterative RTM procedure on velocity correction. The synthetic dataset derived from a two-dimensional (2D) brain model and real datasets acquired from laboratory experiments by Xu and Wang [9] (hereinafter referred as XW06) on rhesus monkey heads will be used for test case. The comparison of imaging results derived by both methods supports the finding of previous studies that RTM is superior to back-projection in imaging quality and accuracy. Meanwhile by analysis the imaging results derived by this iterative RTM procedure, conclusion shows that iterative RTM procedure should be a promising approach for achieving transcranial TAT.

This paper is organized as follows. Firstly back-projection and RTM are briefly introduced in Section 2. Section 3 illustrates the validation of RTM and back-projection methods through both synthetic data and laboratory data. The iterative RTM procedure will be described in detail and demonstrated with the applications to synthetic data and real laboratory data in Section 4. Section 5 is detailed discussion and analysis on the results coming from Section 3 and Section 4. Finally, in Section 6 we restate the major findings as conclusion.

2. Back-projection and reverse time migration

In this section we briefly introduce back-projection (or named Kirchhoff migration in exploration seismic engineering) and RTM, two different migration approaches used for transcranial imaging. Generally speaking migration is an inversion operation involving rearrangement of seismic or acoustic information elements from time to depth domain so that reflections and diffractions are plotted at their true locations [32]. It plays a function of moving dipping events to its original location and increasing the accuracy of the image. From seismic exploration the seismic wave is generated by the source on the ground to propagate downward the earth. Once it hits the non-continuity along sub-surface, reflection wave is generated and detected by transducers located on the ground [33]. The misfits exist between observed data and sub-surface's real location even sources and receivers are perfectly overlapped. The reasons include: 1) Due to the low frequency and non-infinitesimal width of the input beam, the reflection wave could propagate to multiple direction rather than following the single line; 2) Due to the possible non-horizontal structure owned by sub-surfaces, their locations could be "shifted" to wrong position. As Figure 1 shows, combinations of source and receiver are collocated (black solid dots) along the ground line AB. Suppose the velocity is uniform underground, the slope reflectors along CD is shifted incorrectly to C'D' from the results collected by transducers with wave's travel time unchanged. And therefore, the purpose of migration is to shift C'D' back to CD, and hence the correct structure can be reached.

The migration problem can be transformed as a more straight-forward case when it is applied on TAT imaging. The purpose of migration on TAT imaging is to shift the locations of different acoustic sources to their correct locations. In Figure 2, the shadowed area is the to-be-imaged tissue, in which each point is an independent acoustic source after receiving microwave impulse. The dashed circle surrounded shows the location of receiver array. Suppose the velocity distribution in all 2D space is relatively uniform, and tissue's acoustic wavelet is a single narrow pulse, the possible location of certain source S is constrained on a circle with the center as transducer T_N. Its radius is the distance calculated by the wave propagation time took

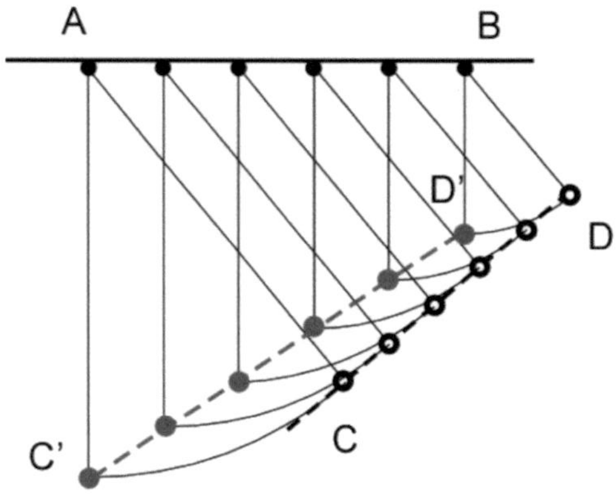

Figure 1. Principle of migration. AB is the ground line; the black solid dots attached are combination of sources and receivers. CD represents dip structures (black hollow dots) underground, it is incorrectly shifted to the location of C'D' (gray solid dots). Migration plays a function of shift structures' coordinate from C'D' back to CD for getting correct result.

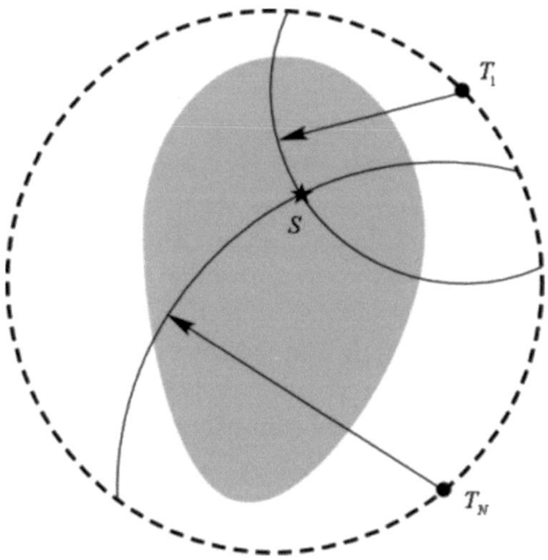

Figure 2. Application of migration on TAT imaging. T_1 and T_N are any two transducers of transducers' array shown as dash circle surrounded with the target tissue (gray shadow). By applying back-projection, the acoustic amplitude of S will be enhanced by multiple transducers.

from S to T_N. During the TAT imaging, for achieving the exact location of S, back-projection and RTM are applied by different principles. Back-projection can be summed as a kind of spatial integration method. It redistributes the received acoustic energies and makes summation in space. In Figure 2, the results observed by different transducers are redistributed along different circles in to-be-imaged domain. If there is a large number of transducers, the amplitude at the location S will be enhanced due to interference of multi circles, otherwise the amplitudes should be minimum [33]. This approach is easily to achieve when the velocity model is uniform; However when the velocity distribution inside the to-be-imaged domain is complex, the possible location of S will be distorted from circle to an irregular shape. Combined with multiple scattering, the assumption of one-way sight transmission in back-projection is no longer established. Different from back-projection, RTM works by running the wave equation backward for all transducers. It sets up all transducers as pseudo sources, and the records are broadcasted with a time-reversed manner-from the end of the trace to time zero. If the velocity model is presumed correctly, in TAT imaging the wave field at time 0 in space will coincide with the original source distribution [19]. Compared with back-projection, by applying full wave equation, RTM is capable of involving all wave field phenomenon including diffraction, aberration and multiple scattering. However it requires much higher usage of CPU time and memory [27].

3. Validation of RTM and back-projection methods

3.1. Validation via synthetic data

To test the effectiveness of back-projection and RTM, we have built a 2D synthetic human brain model with an intact skull. In this model, the brain is made of gray matter and white matter [14], and the skull is made of three layers, namely the inner table, diploe, and outer table [20]. To mimic a real laboratory experiment similar to [8], [13], we modeled the space outside the skull as mineral oil. The reason for us using mineral oil is that it provides both back-ground with uniform acoustic speed and tiny loss coupling between acoustic transducers and human head surface. Additionally when compared with water, mineral oil has much weaker microwave absorption ratio (~0), this will guarantee the minimum errors are introduced during modeling. Instead of mineral oil, a plenty kinds of fluid with weak microwave absorption ratio and comparable acoustic velocity to bio-tissue could also been used in further realistic test. To mimic pathological changes and build a benchmark for results analysis, in the synthetic model we replaced a small area of the brain with blood. This area was located at the left cerebral hemisphere and defined as elliptically-shaped. The distribution of acoustic velocity in the synthetic model is shown in Figure 3a. From a review of several literatures, mechanical parameters of all related bio-tissues were collected and are listed in Table I, in which the velocities and densities of grey matter and white matter were measured from lamb brain using acoustic frequency of 1 MHz [14]. The skull's velocities, densities and thicknesses of different layers are applied from datasets in research [20], [34] and [35]. The loss factor is defined in [35] to depict the amplitude of the propagating acoustic wave's energy decay. Its values come from [14] for brain and [35] for skull, respectively.

After establishing the mechanical properties, we calculated the initial acoustic amplitudes in the synthetic model. We assumed that the initial acoustic pressure (dyne/cm2) in mineral oil and the skull was 0, so the acoustic wave field was entirely generated by multiple acoustic sources in the brain at time zero. Their amplitudes were closely linked to the microwave absorption ratio. As expressed in [6], the relationship between the power intensity I (W/cm2) of absorbed microwaves and the generated peak acoustic pressure P0 (dyne/cm2) is shown as

Parameter Bio-Material	Density kg/m³)	Acoustic speed (m/s)	Loss factor	Thickness (mm)
Grey matter	1039	1483	0.0046	-
White matter	1044	1521	0.0069	-
Outer Table	1870	2900	0.1542	2.0
Diploe	1740	2500	0.1234	2.5
Inner Table	1910	2900	0.1985	1.7
Blood	1057	1500	0	-
Mineral oil	900	1437	0	-

Table 1. Selected mechanical parameters of human brain and skull

Eq. 1, where c is the sound velocity, β is the volumetric thermo-expansion coefficient, and C_p is heat capacity. The definition of absorbed power intensity I can be expressed as Eq. 2 [8], where E, ϱ, and σ are the maximum amplitude of the radiated electromagnetic field, tissue's density, and electrical conductivity, respectively. By combining Eq. 1 and Eq. 2, Eq. 3 is derived as the theoretical relationship between tissue's electrical conductivity and initial acoustic pressure generated based on the thermo-acoustic effect.

$$P_0 = \frac{c\beta I}{C_p} \tag{1}$$

$$I = \frac{\sigma}{2\rho} |E|^2 \tag{2}$$

$$P_0 = \frac{\beta c \sigma}{2c_{pp}} |E|^2 \tag{3}$$

The volumetric thermo-expansion coefficient β and heat capacity Cp of the brain are nearly uniform: (β=12.3X10^{-5} /℃ [36] and C_p=4160 KJ/m^3 ℃ [37]). The acoustic velocity and density vary little among gray matter, white matter, and blood (Table 1). Consequently the initial acoustic pressure can be well approximated as proportional to the electrical conductivity σ. On the other hand, Table 2 shows that for input microwaves the electrical conductivities of white matter, gray matter, and blood are significantly different [8]. For blood in particular, σcan be more than 1.5 times higher than for the other two kinds of tissue. Using the value of electrical conductivity for microwave of 900 MHz from [8] and Eq. 3, the distribution of initial acoustic pressure is derived as shown in Figure. 3b.

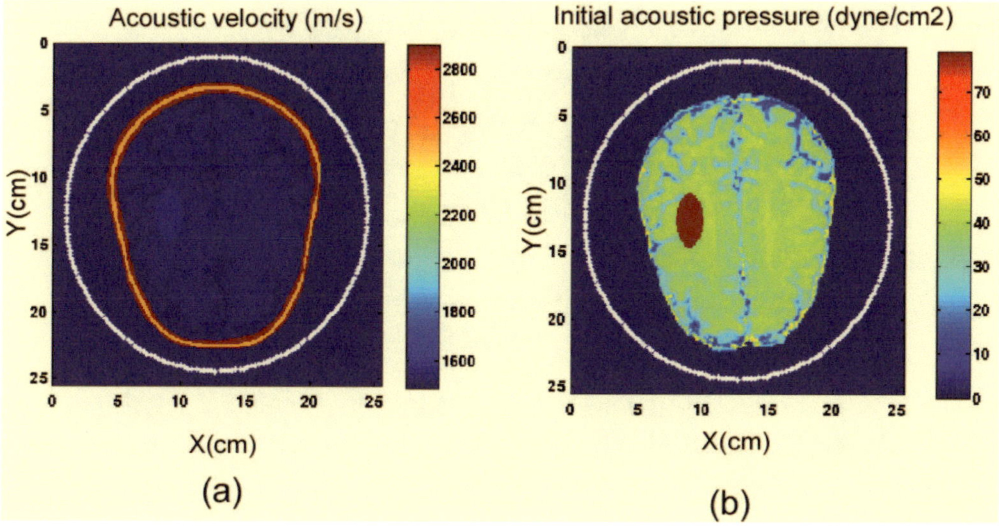

Figure 3. Layout of the 2D human brain model with intact skull, where (a) is the distribution of acoustic velocity in the model, in which the skull exhibits relatively higher velocity; and brain tissue has relatively uniform velocity distribution. The white circle shows the locations of the 240 receivers. (b) shows the amplitude of the initial acoustic pressure stimulated by microwaves during TAT transcranial diagnosis.

	σ(Sm^{-1}) in 900 MHz	σ(Sm^{-1}) in 1800 MHz
White matter	0.665	1.081
Grey matter	1.009	1.525
Blood	1.868	2.283

Table 2. Selected electrical conductivities of human brain

Once the synthetic model was established, we applied the FDTD method to forward modeling acoustic wave propagation. Using the initial acoustic pressure shown in Figure. 3b, a zero-offset Ricker wavelet with a central frequency of 0.15 MHz was applied to each point of the brain as the acoustic source. The model space was meshed as 512 x 512 grids with a spatial interval of 0.5 mm. A total of 1600 time steps with time intervals of 0.115 μs were judged long enough to allow the acoustic waves to propagate to each receiver from the most remote grid in the brain. The outgoing acoustic signal was recorded by 240 receivers located outside the skull (white circle shown in Figure. 3). The derived synthetic time traces are shown in Figure. 4, which provides the input dataset for later imaging.

The velocity model used in both migration algorithms is critical to successful imaging. In this study we applied two velocity models: one (abbreviated as V1) assumes the average acoustic velocity is uniformly 1540 m/s in the model space. Essentially it approximates a "bare brain" model with the effect of the skull excluded. The second model (abbreviated as V2) includes

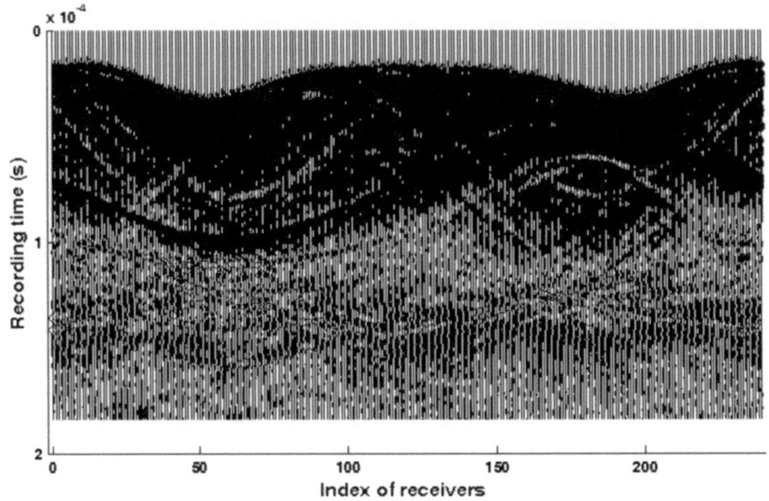

Figure 4. Acoustic signal recorded during FDTD forward modeling of TAT. This is a cluster with 240 traces, with 1600 samples contained in each trace.

the effect of the high acoustic speed of the skull, which is almost double the speed of brain tissues. In our study, velocity models V1 and V2 were applied to both back-projection and RTM. Due to the velocity variance in V2, we applied V2 to two migration methods by different approaches. In back-projection, ray-tracing was applied from every transducer to all directions. This procedure was similar to the methods described in [13], but for simplicity we considered only the rays' travel time caused by velocity variance, and aberration around the skull was ignored. Different from back-projection, as a kind of full-wave migration, RTM uses the same scheme as forward modeling methods such as FDTD. Consequently V2 can be applied to RTM in a straightforward manner. Comparisons of migration imaging results are shown in Figure. 5-7.

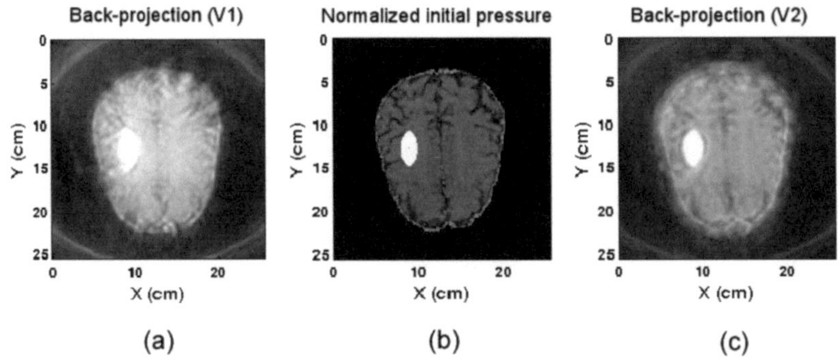

Figure 5. Comparison among back-projection results: (a) shows the results using velocity model V1. (b) shows the initial acoustic pressure of the original model, and (c).shows the back-projection result using velocity model V2

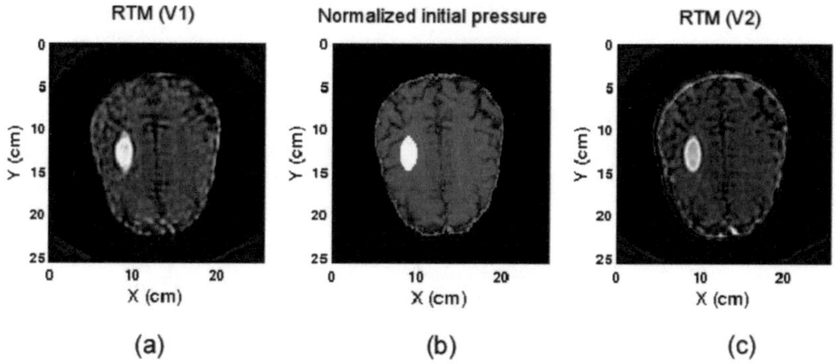

Figure 6. Comparison among RTM results: (a) shows the results using velocity model V1, (b) shows the initial acoustic pressure of the original model, and (c) shows the RTM result using velocity model V2.

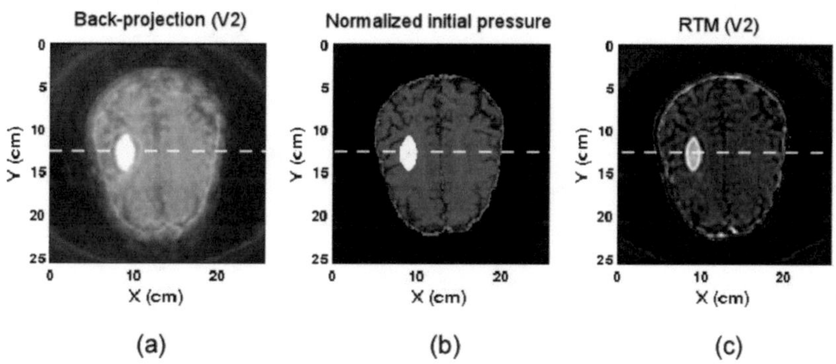

Figure 7. Comparison among KM results: (a), shows the results using velocity model V2 (b) shows the initial acoustic pressure of the original model, and (c)shows the RTM result using velocity model V2.

Figure 5a andFigure 6a are results derived by back-projection and RTM using the "bare brain" model V1. When compared with the distribution of initial acoustic pressure of the original model (Figure 5b or Figure 6b) we can clearly observe two kinds of imaging artifacts. First, the area with higher initial pressure at the left cerebral hemisphere, which is set up artificially in an elliptical-shape, is seriously enlarged by back-projection (Figure 5a) and falsely elongated along the major axis of the ellipse by RTM (Figure 6a); Second, delicate features such as the gyrus, located in the outer part of the brain, and the gap which separates the left and right cerebral hemispheres in our model are totally blurred in both results when using velocity model V1. In contrast, from Figure 5c and Figure 6c, which are results based on velocity model V2 with the skull's velocity included, these two misfits are substantially reduced. From all of these comparisons we can see that exclusion of the skull leads to severe error and distortion in migration imaging for both back-projection and RTM.

To further examine the differences between back-projection and RTM, we reorganized the results using back-projection, RTM, along with the original velocity model V2 as shown in Figure 7. From it we can see that although both back-projection and RTM can transform most of the wave field back to its original location correctly, there are obvious differences in imaging quality between the two methods. Compared with the original model shown in Fig. 5b, the detail features are blurred in back-projection result (Figure 7a) but appear to be clean and sharp in RTM result (Figure 7c). These visual differences can be further amplified through 1-D comparison along the x-direction along the horizontal white dashed line shown in Figure 7 chosen to cross the brain's left boundary, artificially blooded area, interhemispheric fissure, and right boundary. The source amplitudes along this profile for back-projection RTM, and the original model are shown in Figure 8. It is obvious that the result of RTM is far more superior than that of back-projection. Compared with the back-projection result (the dotted line), result of RTM (the solid line) has larger variance for depicting structures such as interhemispheric fissure around 12.50 cm as well as the boundaries of the artificially blooded area around 7.5 cm and 10 cm. These differences can also be clearly observed on brain boundaries around 5 cm and 20 cm, where the amplitude from RTM decays as sharp as the original model, but back-projection's result is obviously incorrect and decay much slower outside of the brain.

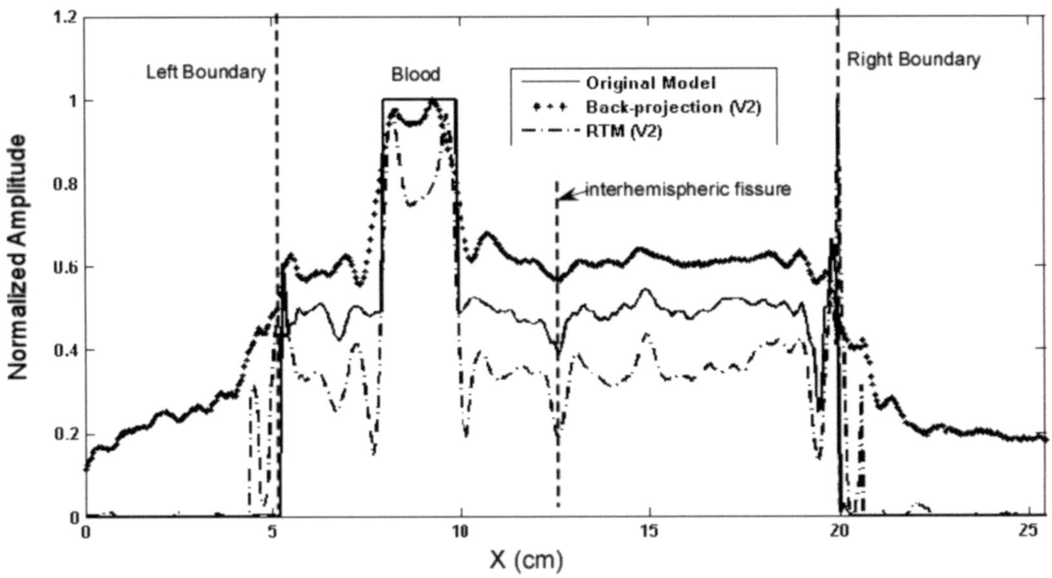

Figure 8. The cross-sections along the white dashed-line shown in Fig.5a-c, which passes brain's left boundary, artificially blooded area, interhemispheric fissure and the right boundary. The dotted-broken line, dotted line, and solid line show the original model (Fig. 5b), the results by back-projection (Fig. 5a), and RTM (Fig. 5c) using the model V2.

3.2. Validation through laboratory data

The back-projection and RTM algorithms were also tested by using the laboratory data acquired by XW06. In their experiment, the monkey's head was decapitated and fixed by a clamp and completely immersed in mineral oil. During TAT detection, this specimen was stimulated by 3-GHz microwave pulses, and the derived acoustic wave field was recorded by a transducer with a 1 MHz central frequency and about 0.8 MHz bandwidth. The transducer was positioned from 6-14 cm to the center of the monkey's head, and the sampling frequency was 20 MHz. During the experiment, the clamp fixing the monkey's head was mounted on a rotary table driven by a stepper motor with a step size of 2.25 degrees. Accordingly in this laboratory application, the outgoing acoustic wave was observed by 160 receivers surrounding the head in a 2D circle. With data processing performed through the procedure of [38] for high frequency enhancement, only the segment with a spectrum of 0.3-1 MHz of the observed data was picked up and enhanced for imaging by back-projection and RTM. We applied the estimated average acoustic velocity to both image approaches, since any velocity information on velocity distribution of earlier experiment in XW06 was unknown, which may introduce some error and reduce the image quality.

Figure 9 shows the results based on a dataset collected from a one-month-old monkey head with a skull thickness of less than 1 mm. The velocity model is assumed to be uniform as 1437 m/s, as the same as acoustic velocity of mineral oil. The region shown is 53mm by 51mm along the coronal cross section. From the experiment of XW06, three steel needles with diameters of 0.9 mm were inserted in the approximate locations as shown in Figure. 9a (XW06). The results derived from back-projection (Figure. 9b) and RTM (Figure. 9c) are shown side-by-side for comparison. Both imaging algorithms show the three needles, and the black dot located at the center is believed to be an air bubble introduced by inserting the needles (XW06). Compared with the result derived from back-projection, the needle A in RTM result owns sharper edges. Meanwhile, back-projection provides less visibility for needle C than RTM. From the plots along the x cross-section shown in Figure. 9d, although both needle A and B can be detected by using back-projection and RTM, the back-projection image is much nosier. The existence of this noise causes seriously reduction of signal noise ratio and fussy in whole image by back-projection. However needle B is seriously distorted from results of both back-projection and RTM. This shows that due to the technical limitations of the coarsely estimated average acoustic velocity for TAT reconstruction, neither of these two methods can provide satisfying image quality.

4. Iterative RTM procedure

The previous result shows that the image quality of RTM is restricted by the precision of the velocity model. In some sense, the increased requirement of RTM sensitivity to indistinct velocity model brings a paradox: If the velocity model is well known by operation or any invaded techniques, RTM is not necessary; If the velocity model is not known, RTM lacks an essential input. For solving this paradox and improving the RTM result, rather than single RTM run, we developed iterative RTM procedure based on the theory of Whitmore [23]. The iterative RTM procedure works based on an assumption between velocity model and RTM result. Suppose there exists a function to map velocity from RTM result, complex velocity model could be renewed after RTM for one time. By repeating this procedure, velocity model will continue be updated after multiple RTM runs until the result becomes promising. This iterative RTM procedure is shown in Figure 10 as a schematic. An initial guess of a model is

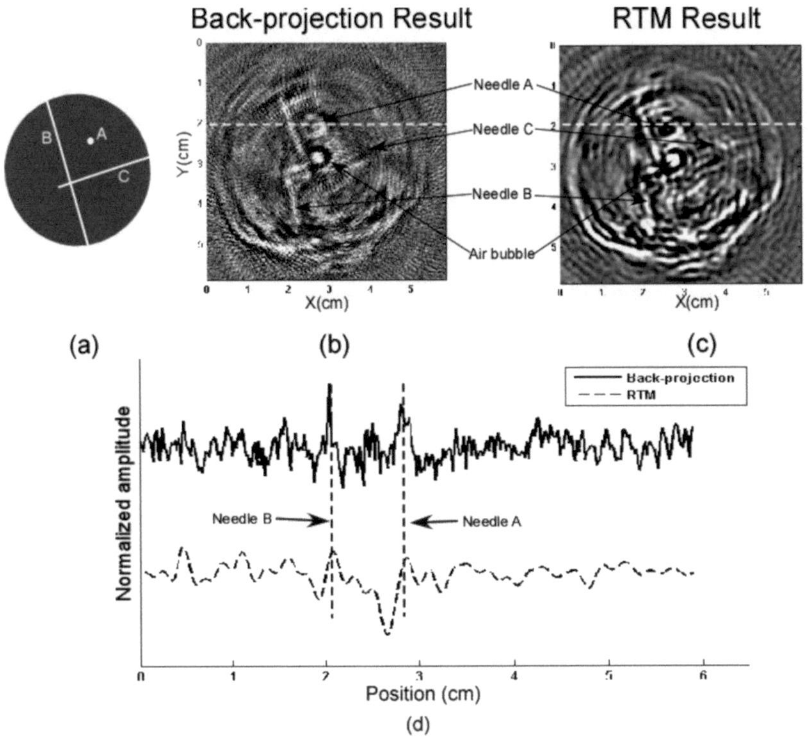

Figure 9. a) Diagram showing a monkey head with three inserted needles from XW06; (b) TAT result derived by KM; (c) TAT result derived by RTM; (d) Line plot along the white dashed line at 2 cm of (b).

made, incorporating all known external velocity information. The image is derived through RTM. After that a new velocity model is made based on current RTM result. This procedure is repeated until both RTM results and velocity models are unchangeable.

As Figure 10 shows, the key procedure of iterative RTM is to update velocity model after each RTM. This requires a function for mapping velocity from image derived by RTM. Since for TAT imaging RTM image is the reconstruction of electrical conductivity distribution, the desired function of velocity and electrical conductivity could be built by referring to Table 1 and Table 2. Based on the given velocity and electrical conductivity values of white matter and gray matter, a function is established by using the second order polynomial fitting as shown in Figure 11. Obviously this function is coarsely approximated. It could be improved by involving more velocity–electrical conductivity pairs in further research. Meanwhile it's noteworthy that the mapping function can't derive correct velocity on blood, water and mineral oil, consequently a mask must be used to preserve the region outside sample from wrongly updating. At present the function shown in Figure 11 is used in our research to map the velocity from electrical conductivity derived by RTM; in this way the velocity model will be updated for the next RTM iteration.

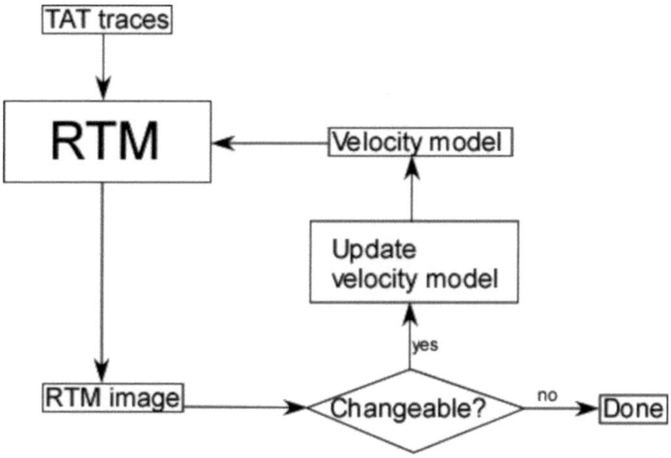

Figure 10. Schematic of iterative RTM procedure

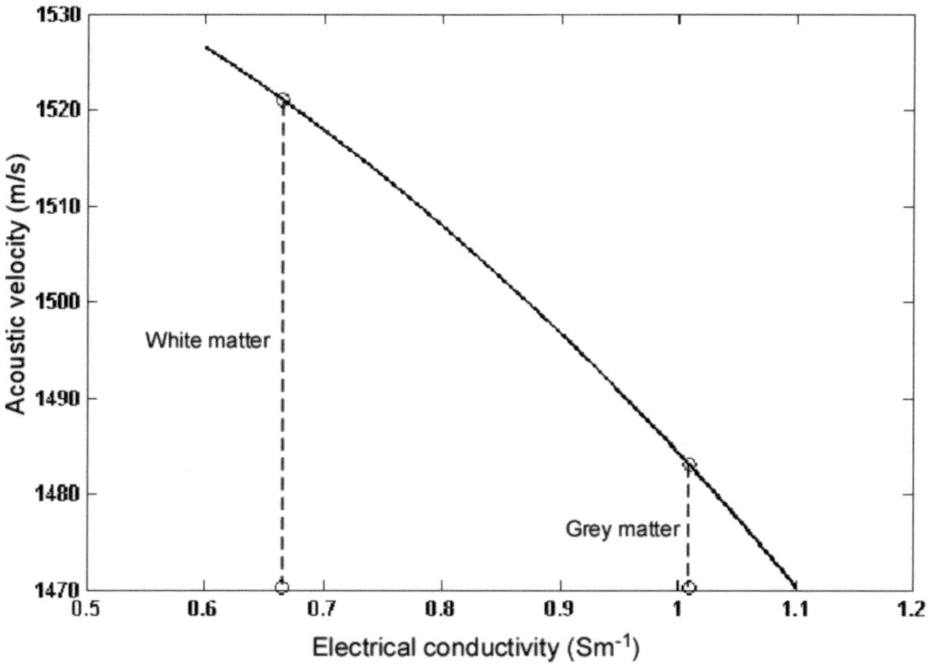

Figure 11. Mapping function for calculating acoustic velocity through RTM derived electrical conductivity. This function is second-order polynomial fitted by velocity-electrical conductivity pairs of white matter and grey matter given by Table 1 and Table 2.

An example of applying iterative RTM procedure is shown in Figure 12. The laboratory data of this case comes from XW06, which has been introduced in Section V. By referring to XW06, the velocity model would not be known during TAT. The only knowledge of the velocity field in this case is 1437 m/s of mineral oil, which submerges the sample during TAT detection. To demonstrate the iterative RTM procedure, we make an initial guess that velocity model is uniform as 1437 m/s. By using the initial velocity model, the first RTM output is derived as shown in Figure 12b, which is as the same as Figure 9c. Velocity model will then be updated as new velocity input for RTM. This procedure is repeated for several times until RTM output is unchangeable. The TAT results derived from iterative RTM procedure on step 3 and 5 are displayed on Figure 12b and Figure 12c.

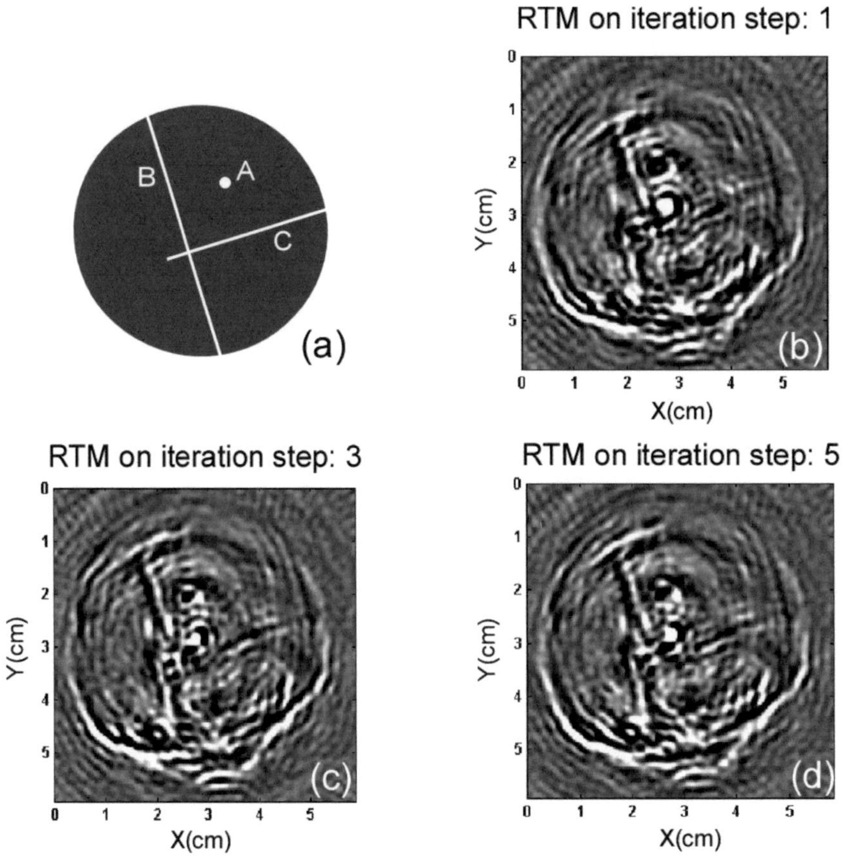

Figure 12. a) Diagram showing a monkey head with three inserted needles from XW06; (b) TAT result derived from iterative RTM procedure on step 1; (c) TAT result derived from iterative RTM procedure on step 3; (d) TAT result derived from iterative RTM procedure on step 5.

From Figure 12, by using iterative RTM procedure the improvement on TAT image can be observed from several aspects: First, by comparing with Figure 12b (Figure 9c), the distortion of needle B has been well corrected after five times iteration. Second, the original RTM result shown in Figure 12b provides less visibility for needle C. After two iterations, Figure 12c shows that the boundary of needle C has been largely enhanced. Additionally, comparing among Figure 12 a-c, the cross-section of needle A has been focused gradually with iteration times

http://dx.doi.org/10.5772/56619

increasing. All of this shows that iterative RTM procedure has great potential of correcting coarsely estimated velocity model and therefore enhance imaging capability of RTM.

5. Discussion

By combining Section II and III, It is clear that RTM is superior to back-projection in terms of imaging quality and higher signal to noise ratio. Compared with back-projection, which makes a ray approximation, RTM bases its entire algorithm on solving a full-wave equation, without substantial approximation, and holds the original dynamic features of the wave field intact. The handle of velocity heterogeneity is fundamentally intrinsic. Usually, adapting ray tracing in back-projection is time consuming and has a limited improvement on image quality. Unlike back-projection, the quality of RTM's results is independent of the complexity of the velocity model. This feature makes the wave propagation in the domain highly accurate compared with using ray-tracing. Figure 5-8 show that RTM is able to recover almost all features of the brain to their original potion when the skull's velocity is included.

RTM can recover a complex structure's boundary sharply. This has been reported by [19], [27] and is proved by our results in Figure 6. When both velocity models V1 and V2 are used in RTM, the boundaries of tiny features can be clearly seen. Especially, even though obvious distortions exist in the result using V1 (Figure 6a), with the exclusion of the skull, all features are still relatively un-blurred in comparison with the back-projection results (Figure 5a). By comparison, looking at the cross-section in Figure 8, the sharp edges of brain are well recovered by RTM but seriously smeared by back-projection. Further, in Figure 9 when the skull-excluded model is applied, the contour of Needle C is well recovered by RTM but not by back-projection.

Nevertheless, it is noteworthy that the image quality of RTM is still limited by the precision of the velocity model. Consequently the key to capitalizing on the benefit of RTM is to build better velocity models before applying RTM. For achieving this goal, we developed iterative RTM procedure to update velocity model iteratively. Its principle is based on an assumption that acoustic velocity could be mapped from RTM image through a certain function. The results shown in Figure 12 demonstrated that currently even by using our coarsely approximated velocity function, the quality of RTM image can be well improved after several iterations within iterative RTM procedure. It could be improved by involving more velocity–electrical conductivity pairs in our further research.

6. Conclusion

In this paper we have compared back-projection and RTM for transcranial TAT imaging. Compared with back-projection, RTM offers better performance with regard to velocity variance, imaging quality, and noise suppression caused by spatial aliasing. The capability of RTM can be further improved by iterative RTM procedure, which aimed to provide velocity model with higher accuracy. Generally speaking RTM owns great potential for achieving acoustic localization on transcranial TAT imaging with high quality and accuracy.

Author details

Zijian Liu* and Lanbo Liu

*Address all correspondence to: lzjknight34305@gmail.com

School of Engineering, University of Connecticut, Storrs, CT, USA

References

[1] M. Xu and L. V. Wang, "Time-domain reconstruction for thermoacoustic tomography in a spherical geometry," *IEEE Trans. Med. Imag.*, vol. 21, pp. 814-822, Jul. 2002.

[2] Y. Xu, D. Feng and L. V. Wang, "Exact frequency-domain reconstruction for thermoacoustic tomography-I: Planar Geometry," *IEEE Trans. Med. Imag.*, vol. 21, pp. 823-828, July 2002.

[3] Y. Xu, D. Feng and L. V. Wang, "Exact frequency-domain reconstruction for thermoacoustic tomography-II: Cylindrical Geometry," *IEEE Trans. Med. Imag.*, vol. 21, pp. 829-833, July 2002.

[4] W. Joines, R. Jirtle, M. Rafal, and D. Schaeffer, "Microwave power absorption differences between normal and malignant tissue," *Radiation Oncol. Biol. Phys.*, vol. 6, pp. 681-687, 1980.

[5] M. S. Hawley, A. Broquetas, L. Jofre, J. C. Bolomey, and G. Gaboriaud, "Microwave imaging of tissue blood content changes," *J. Biomed. Eng.*, vol. 13, pp. 37-44, 1991.

[6] K. R. Foster and E. D. Finch, "Microwave hearing: evidence for thermoacoustic auditory stimulation by pulsed microwaves," *Science*, vol. 185, pp. 256-258, 1974.

[7] R. Nikolaus, "Thermoacoustics," *Adv. Appl. Math.*, vol. 20, pp. 135-175, 1980.

[8] M. Martinez-Burdalo, A. Martin, M. Anguiado and R. Villar, "Comparison of FDTD-calculated specific absorption rate in adults and children when using a mobile phone at 900 and 1800 MHz," *Phys. Med. Biol.*, vol. 49, pp. 345-354, 2004.

[9] Y. Xu and L. V. Wang, "Rhesus monkey brain imaging through intact skull with thermoacoustic tomography," *IEEE Trans. Ultrason. Ferroelectr. Freq. Control.*, vol. 53, pp. 542-548, Mar., 2006.

[10] D. Royer, "Mixed matrix formulation for the analysis of laser-generated acoustic waves by a thermoelastic line source," *Ultras. Internat.*, vol. 39, pp 345-354, 2001.

[11] N.H. Noroy, D. Royer, M. Fink, "Transient elastic wave generation by an array of thermoelastic sources," *Appl. Phys. Lett*, Vol.63, pp. 3276-3278, Dec 1993.

[12] J. Gamelin, A. Aguirre, A. Maurudis, F. Huang, D. Castillo and Q. Zhu, "Curved array photoacoustic tomography system for small animal imaging," *J. Biomed. Opt.*, vol. 13, no.2, Mar.-Apr., 2008.

[13] X. Jin, C. Li and L. V. Wang, "Effects of acoustic heterogeneities on transcranial brain imaging with microwave-induced thermoacoustic tomography," *Med. Phys.*, vol. 35, no. 7, 2008.

[14] S-C. Lin, S-J. Shieh and M. J. Grimm, "Ultrasonic measurements of brain tissue properties," *Center for Disease Control conf. Pre-proceedings*, pp. 27-31,1997.

[15] W. A. Schneider, "Developments in seismic data processing analysis," *Geophysics*, vol. 36, pp 1043-1073, 1971.

[16] W. A. Schneider, "Integral formulation for migration in two and three dimensions," *Geophysics*, vol. 43, pp 49-76, 1978.

[17] W. S. French, "Computer migration of oblique seismic reflection profiles," *Geophysics*, vol. 51, pp. 961-980, 1975.

[18] A. J. Berkout, *Seismic migration imaging of acoustic energy by wave field extrapolation*, Elseiver Science Publ. Co., Inc., 1982.

[19] J. Zhu and L. R. Lines, "Comparison of Kirchhoff and reverse time migration methods with applications to prestack depth imaging of complex structures," *Geophysics*, vol. 63, pp. 1166-1176, 1998.

[20] F. J. Fry and J. E. Barger, "Acoustic Properties of Human Skull," *J. Acoust. Soc. Am.*, vol. 63, pp. 1576-1590, 1978.

[21] C. Hemon, "Equations d'onde et modeles," *Geophys. Prosp*, vol. 26, pp. 790-821, 1978.

[22] G. A. McMechan, "Migration by extrapolation of time-dependent boundary values," *Geophys. Prosp.*, vol.31, pp. 412-420, 1983.

[23] N. D. Whitmore, "Iterative depth migration by backward time propogation," *53rd Ann. Internat. Mtg. Of Soc. Expl. Geophys.*, Expanded abstracts, pp. 827-830, 1983.

[24] E. Baysal, D. D. Kosloff and J. W. C. Sherwood, "Reverse-time migration," *Geophysics*, vol. 48, pp. 1514-1524, 1983.

[25] K. Yee, "Numerical solution of initial boundary value problems involving Maxwell's equations in isotropic media," *IEEE Trans. Antennas Propag.*, vol. 14, 1966.

[26] Liu, Q. H., "The PSTD algorithm: A time-domain method requiring only two cells per wavelength," *Microw. Opt. Technol. Lett.*, Vol. 15, No. 3, Jun. 1997.

[27] P. Farmer, Z. Zheng and D. Jones, "The role of reverse time migration in imaging and model estimation," *The leading edge, Soc. Expl. Geophys*, vol. 28, pp. 436-441, 2009.

[28] Z. Liu and L. Liu, "Transcranial Thermoacoustic Tomography: A Comparison of Two Imaging Algorithms," *IEEE Trans. Med. Imag.*, Vol 32, pp. 289-294, 2013.

[29] X. Jin and L. V. Wang, "Thermoacousitc tomography with correction for acoustic speed variations," *Phys. Med. Biol.*, vol. 51, pp. 6437-6448, 2006.

[30] Z. Liu and L. Liu, "A Novel Approach for Thermoacoustic Tomography by Kirchhoff Migration," *9th International Conference on Theoretical and Computational Acoustics*, Germany, 2010.

[31] L. Liu, K. He and L. V. Wang, "Transcranial ultrasonic wave propagation simulation: skull insertion loss and recovery," in *SPIE conf.*, San Francisco, USA, 2007.

[32] R. E. Sheriff, *Encyclopedic dictionary of applied geophysics*, fourth edition, Society of Exploration Geophysicists, 2002.

[33] R. E. Sheriff and L. P. Geldart, *Exploration seismology*, second edition, Cambridge University Press, 1995.

[34] N. Lynnerup, J. G. Astrup and B. Astrup, "Thickness of the human cranial diploe in relation to age, sex and general body build," *Head and Face Medicine*, 1-13, 2005.

[35] M. Hayner and K. Hynynen, "Numerical analysis of ultrasonic transmission and absorption of oblique plane waves through a human skull," *J. Acoust. Soc. Am.*, vol. 110, pp. 3319-3330, Dec., 2001.

[36] J. C. Lin, "Microwave-induced hearing: some preliminary theoretical observations," *J. Micro. Power*, vol. 11, No. 3, 1976.

[37] X. Xu, P. Tikuisis and G. Giesbrecht, "A mathematical model for human brain cooling during cold-water near-drowning," *J Appl Physiol*, vol. 86, pp. 265-272, 1999.

[38] Y. Xu and L. V. Wang, "Signal processing in scanning thermoacoustic tomography in biological tissues," *Med. Phys.*, vol. 28, pp. 1519-1524, Jul. 2001.

Chapter 4

Automatic Identification and Interpretation of Animal Sounds, Application to Livestock Production Optimisation

Vasileios Exadaktylos, Mitchell Silva and
Daniel Berckmans

Additional information is available at the end of the chapter

http://dx.doi.org/10.5772/56040

1. Introduction

In modern livestock production, biosecurity and improved disease monitoring are of great importance to safeguard public and animal health. In addition, citizens expect that farm animals have been reared and killed humanely with minimal environmental impact and that food from animals is safe. These global threats and concerns can be combated through surveillance and research networks for early detection of animal diseases and better scientific cooperation between countries and research teams. The use of information technology (IT) can provide new possibilities for continuous automatic monitoring (Fig. 1) and management of livestock farming, according to multiple objectives.

Surveillance relies on vast streams of data to identify and manage risks. Currently data collection nearly always depends on manual methods. While this might be acceptable in R&D projects, it is unrealistic when solutions are applied on the scale needed on commercial farms. Scoring of some animal-based information by human inspectors by manual methods is often inaccurate, time consuming and expensive when implemented at farm level. It is clear that a multidisciplinary and integrated approach, using modern IT systems, is needed to optimise use of expensive inspectors.

One current advantage provided by technology in livestock production is the development of sensors and sensing technologies to automatically monitor and evaluate data from processes in real-time. Collecting data from livestock and their environment is now possible using innovative, simple, low-cost IT systems; data are then integrated in real time by using knowledge-based computer models [1].

Figure 1. Sound analysis can be used in a variety of applications. Pigs, chickens and chicken embryos are considered in this chapter.

The initial application of technology in livestock has been the growth of housed pigs and poultry though, in principle, this approach could be applied to any farmed species, including animals farmed extensively [2]. Stockmen routinely gather auditory, olfactory and visual information from their animals to evaluate health, welfare and productivity. New technology can aid this task, even with large flocks or herds, thanks to the (r)evolution in sensors and sensing techniques, e.g. developments in micro- and nano- electronics [3][4].

Other sensors include pedometers for monitoring oestrus behaviour in dairy cows [5]. Automatic weighing systems for broilers, laying hens and turkeys have been used for a number of years to estimate the average weight of a flock [6]. Telemetry sensors for measuring heart rate, body temperature and activity have also been developed [7]. Sensors for quantifying milk conductivity and yield of individual cows are available and may be used to optimise production and provide early detection of poor welfare in individuals [8]. The above examples are not exhaustive, but demonstrate the present and future possibilities in monitoring animal disease, welfare and performance.

The last decade a new field of research has appeared in relation to livestock production and animal welfare. Precision Livestock Farming (PLF) has emerged as a tool that uses continuous and automatic techniques in real-time in order to monitor and control/manage animal production, health, welfare and environmental load in livestock production. The first step in PLF is the measurement and interpretation of animal bioresponses. In this direction, sound has been extensively used as an important bioreponse that can provide useful information regarding the status of the animals.

In this chapter, three examples of monitoring animal sounds as a tool to determine animal status are presented, namely chicken embryos, chickens and pigs.

2. Interpretation of chicken embryo sounds

Industrial egg incubators vary in size and capacity from 10.800 to 129.600 eggs, with the larger machines being commercially more attractive to hatchery managers due to the lower investment and operational costs per egg. It has been shown in the relevant literature that the spread of hatch of the eggs in an industrial incubator can be as long as 48h [9]. This has serious implications to the later life of the chicks because for operational reasons access to feed and water is delayed for the early hatchlings.

To study the effect further, techniques for monitoring the hatch window have been developed. For example, monitoring climate variables (e.g. temperature and humidity) along with biological variables (e.g. heart rate) have shown potential as estimators of the hatching process. More specifically [10] studied the CO_2 balance as an indirect measure of thermoregulation. In [11] heart rate fluctuations were studied and were shown to have specific patterns during the pipping and hatching stages. The resonance frequency of incubating eggs was also linked to hatching and even a predictor of hatching could be made [12]. More recently, it was shown [13] that magnetic resonance imaging (MRI) can be used to monitor hatching.

The above techniques are valuable tools in a research environment. However, their invasive nature and/or high cost of implementation don't allow for use in an industrial or commercial setup. To overcome such problems, sound was used to study the hatching process [14]. Sound can be measured non-invasively and a single microphone can acquire information about the complete incubator which makes it attractive for the industry. At the same time, current technology makes it possible to use microphones in different environment at a low cost.

At the same time, research has been conducted in relation to reducing the hatch window (i.e. the time between the first and last chick that hatched) [15]-[17]. However, successful application of many of these techniques requires accurate identification of Internal Pipping (IP – the moment in which the embryo is entering the air cell inside the egg) which is an important milestone in the hatching process. Once this stage has been reached, chicks inside the egg start to vocalise.

To address this problem of identifying the IP stage, an algorithm has been developed where sound is recorded in an industrial incubator [18], is analysed in real-time using a Digital Signal Processor (DSP) and is able to detect when all the chicks in the incubator have reached the IP stage (i.e. IP100). Despite the fact that the collected sound is coming from the whole incubator, the developed technique is able to isolate the sounds coming from all the eggs and infer information about the embryo state. The algorithm is based on the observation that the frequency content of the embryo sounds during different hatching stages is significantly different [18]. This is visually illustrated in Fig. 2.

2.1. Algorithm description

Although the vocalisations at different stages of the hatching process do show statistically significant differences in terms of the peak frequency, a successful classification algorithm requires a robust feature that will not (or minimally) be affected by other sounds that are acquired by the microphone. To account for this, the peak frequency of a sound is not used directly by the algorithm. From the results of Fig. 2 and by experimentation a ratio of the total energy in the 2-3 and 3-4 kHz frequency bands was able to provide a robust measure that can easily be calibrated as explained below. This difference is visualised in Fig. 3 where the mean spectra of the collected sounds are presented.

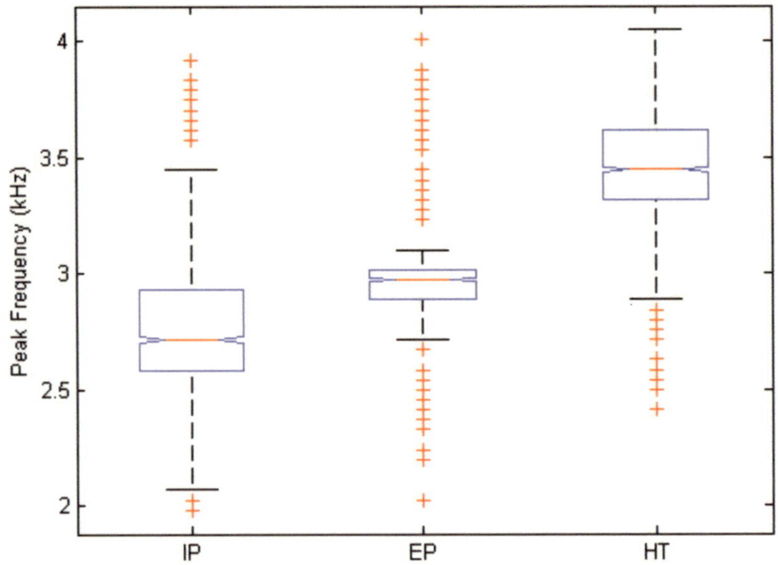

Figure 2. Boxplot showing the peak frequency of the collected sounds when the embryo is in Internal Pipping (IP), External Pipping (EP) and when it has hatched (HT). The box shows the 25th and 75th percentile of the data, the whiskers are the most extreme points not considered outliers and the crosses are the outliers.

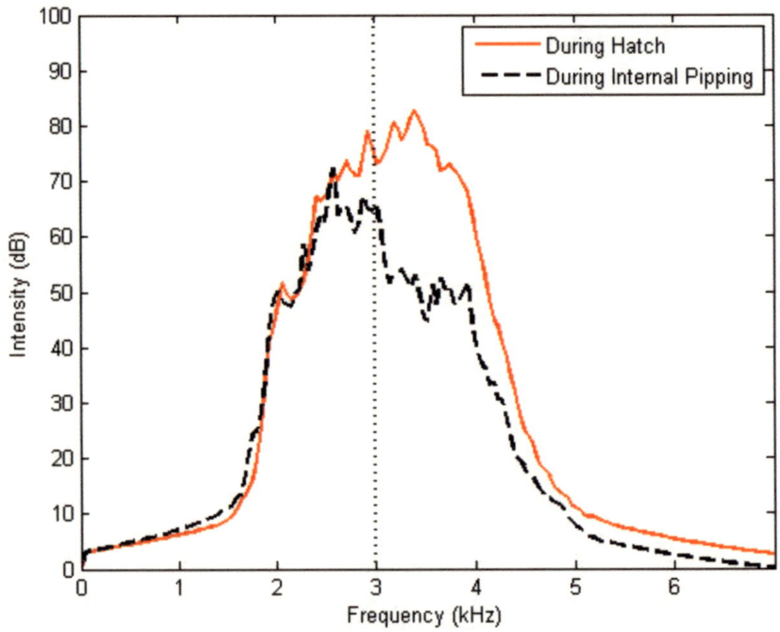

Figure 3. Frequency content of the mean of vocalisations during Internal Pipping (dashed black line) and the mean of vocalisations during Hatch (solid red line). A line visualising the limit of 3 kHz is also shown (vertical blue dotted line)

The classification of the frequency ratio is based on an adaptive threshold. This threshold reflects the background noise characteristics during a period where IP has not started yet. During the beginning of the sound recording (e.g. Incubation Day 16) when the embryos have not penetrated the air cell yet, the algorithm is defining a baseline. The threshold is then set to 90% lower than this threshold. Once the algorithm output remains below this threshold value for more than 5 minutes continuously, IP100 has been reached an the algorithm is giving a signal. In the training dataset, the algorithm has shown to have an accuracy of ±3.1 hrs in the detection of IP100.

A block diagram description of the algorithm is shown in Fig. 4.

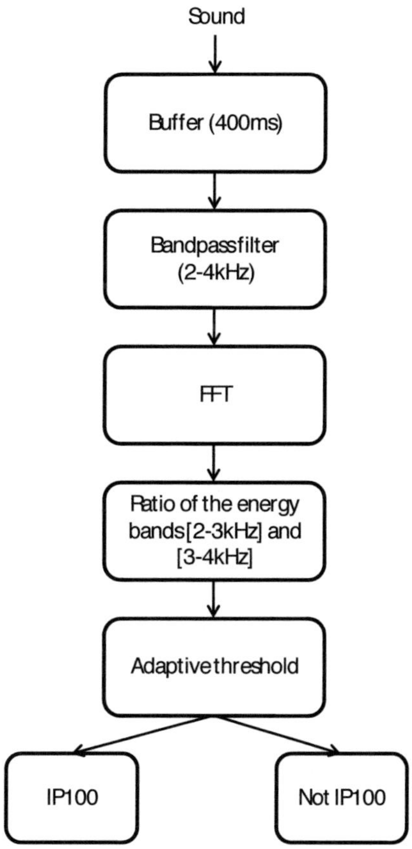

Figure 4. Block diagram of the algorithm for automatic identification of IP100

2.2. Validation results

The algorithm was developed and tested using data collected during 12 incubations in an industrial incubator (Petersime, Belgium). The algorithm had an absolute deviation of 3.2 hrs from the actual IP100 point. An example of the output of the algorithm is presented in Fig. 5.

As explained in the previous section, the threshold shown in Fig. 5 is automatically defined as 90% of the mean value during the initialisation of the algorithm. This value was chosen after experimentation with the data and allows for short variations of the algorithm output due to non-incubation sounds not to be considered as an alarm.

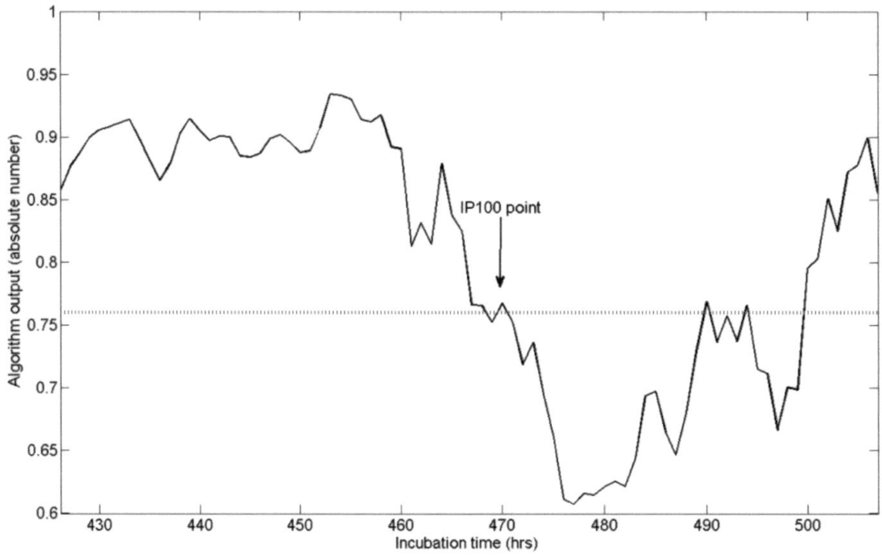

Figure 5. The output of the algorithm (black solid line) and the threshold that was automatically chosen (horizontal blue dotted line). Once the algorithm output crosses the threshold for more than 5 minutes, IP100 is detected.

For the validation, the algorithm was implemented on a DSP (TMS320C6416T by Texas Instruments) using the Real-Time Workshop® of the MATLAB® environment. Once it was giving the signal that IP100 has been reached, a sample of 300 eggs was manually checked for IP. This experimental setup is shown in Fig. 6. In total, 5 experiments were performed during 5 incubations and the results are shown in Table 1.

Trial no.	Time of IP100 detected by the algorithm (incubation time in h)	Manual count of IP (%)
1	467	93
2	468	97
3	470	96
4	469	96
5	470	98

Table 1. Validation results of the algorithm for IP100 detection in industrial incubators

The results of Table 1 show that IP100 was identify correctly. Although this may seem an inaccurate claim (since the IP count was between 93% and 98%), it is important to note the nature of the validation trial. Once the algorithm is providing a signal, IP is manually checked in the incubator. In the hypothetical case that IP is shown to be 100% (i.e. all embryos have been through the IP stage), it is not certain when this has happened. It is perfectly possible that IP100 has been reached long before the manual count. However, if IP of 98% is manually checked, then this means that almost all the embryos have been in the IP stage and therefore from a practical point of view, it is very close to the actual IP100.

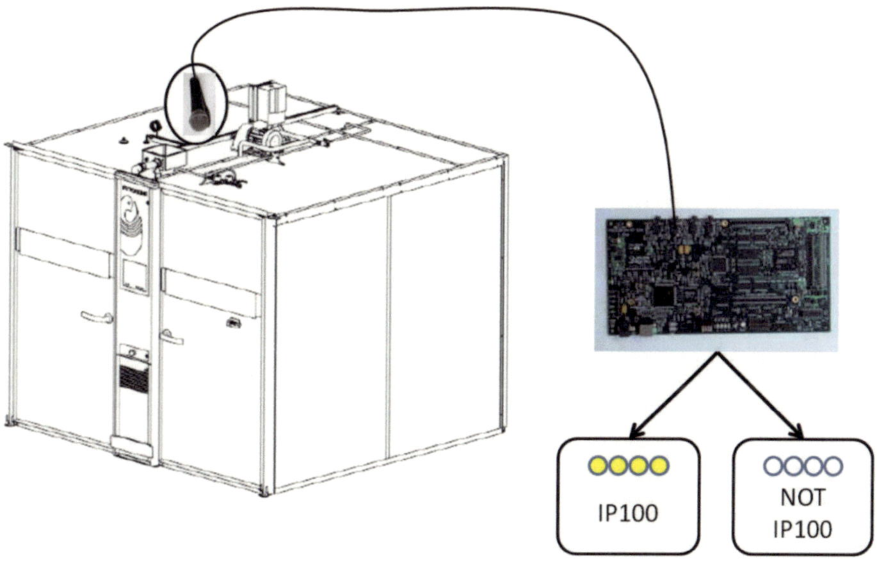

Figure 6. Schematic diagram of the experimental setup for the validation trial. The sound is acquired from a microphone placed in the left part of the incubator and the signal is directed to a Digital Signal Processor (DSP). The DSP is executing the algorithm and is providing a signal by means of LEDs as to whether IP100 has been achieved.

3. Monitoring of vocalisations of chickens

Analysis and interpretation of vocalisations of chickens is a powerful tool when studying the behaviour and welfare. It has been presented in the literature that different types of calls (such as distress or happy calls) can be identified based on their frequency characteristics [19], [20]. More recently, attempts are made to automatically classify these calls and relate them to animal welfare [21].

In relation to thermoregulation, chicken vocalisations have been shown to have specific patterns when chicks are exposed to cold stress [22]. Furthermore, vocal solicitation of heat, forms an integral part of the development of the thermoregulatory system of young chickens [23]. Finally, the thermal comfort of chicks has been shown to have links with the amplitude and frequency of calls [24].

On a social level, chick vocalisations are directly linked to stress and welfare parameters. A dissociation of stress behaviour during social-separation of the chick is mediated by novelty while distress calls are mediated by isolation [25]. Lastly, a method has been developed that could be used for stress monitoring of a flock and is based on multi-parametric sound analysis [26].

It should be noted that chickens, being living organisms, are Complex, Individually different, Time varying and Dynamic (CITD) [27]. As such, the characteristics of their vocalisations are expected to change with time. This is confirmed by experimental results in which the peak frequency of the chicken vocalisations for the same chicken at 1 day and at 20 days old was estimated to be 3603(±303) Hz and 3441(±289) Hz respectively. This parameters shows to be significantly different between the two groups.

It should be noted that no distinction was made between different types of vocalisations. For example, in [19] different types of calls (e.g. distress or pleasure calls) were studied and it was shown that there are differences in terms of frequency content. A more detailed analysis is needed in terms of the evolution of the frequency characteristics with age.

4. Pig cough identification for health monitoring

Similar to the effect on humans, respiratory diseases in pigs result in coughing and in a different sound of coughing due to the different response of the respiratory system when contacting different pathogenic agents. In humans, an experienced physician can identify over 100 different respiratory diseases based on the sound timbre [28]. In animals, veterinarians use a similar approach to detect sick animals when they enter a farm. Their initial impression over the herd is based on visual and auditory observation when they collect information about the welfare, health and productive status of the animals. In this direction, pig vocalisations related to pain were studied in [29], while vocalisation analysis in livestock farms as a measure of welfare has been employed in [30], [31].

The frequency characteristics of pig vocalisation have been extensively studied under different conditions [32][33][34]. These approaches have been further extended to develop algorithms for automatic classification of pig coughing under commercial farming conditions. For this, both frequency [35] and time domain [36] sound analysis techniques have been used. Furthermore, the distinction between dry and productive coughs has been made to further enhance the capabilities of the algorithm. This latter has been studied in detail in [37] where the energy envelope of a sound was shown to be different in dry and productive cough. The system is therefore able to not only provide information about the number of coughing incidents, but also in relation to the quality of the produced cough.

4.1. Algorithm description

As mentioned above, the objective of the algorithm is to detect productive and non-productive coughs in a commercial piggery. Sound is acquired in the pig house by a single microphone placed in the middle of the compartment at a height of 2 meters above the ground. The overall algorithm is shown in Fig. 7 and each block is subsequently described in more detail.

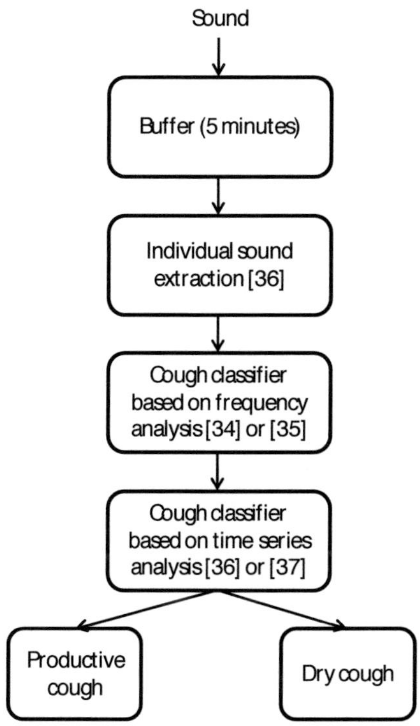

Figure 7. Block diagram of the algorithm to identify cough and distinguish between productive and dry cough.

The first step of the algorithm is to identify individual sounds before using a cough classifier. A procedure that is based on the energy envelope of the sound signal has been described in [36]. Briefly explained, this approach is identifying the parts of the signal with high enough energy that could potentially be cough incidents. An example of this is shown in Fig. 8 for a cough attack (a series of coughing events). It should be noted that each sound is identified as an event for further processing. There is no guarantee that the sound originates from a single animal while overlap of sounds is possible (especially during busy periods of the day).

It can be argued that a such a simple sound extraction algorithm would require accurate calibration of the microphone and would be prone to errors due to large background noise. To tackle this, depending on the situation, noise filters can be included preceding the application of the sound extraction algorithm in order to remove any structured and know background noise. At the same time, an adaptive threshold procedure has also been integrated [36] that allows for accurate selection of only sounds events without including noise. It was further shown in [36] that despite the background noise, classification is still possible using this sound extraction method.

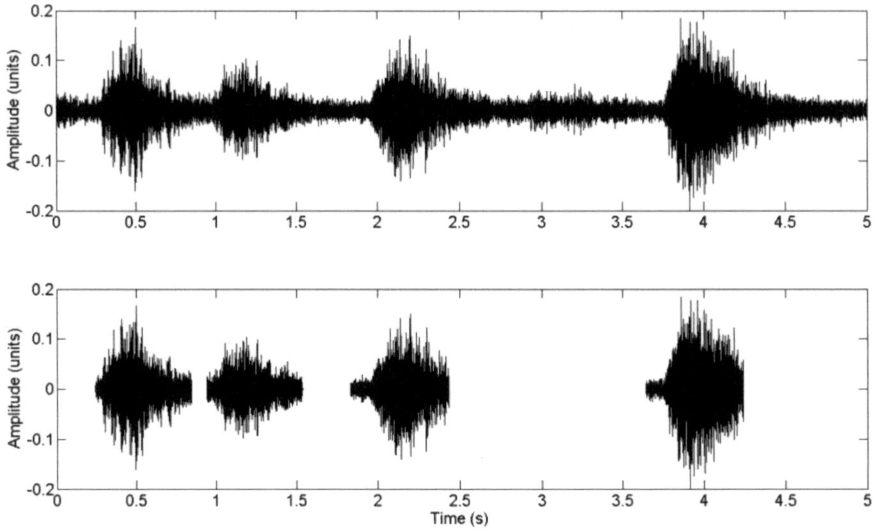

Figure 8. Extraction of individual sounds (bottom plot) from a cough attack (top plot). All 4 sound incidents are successfully extracted.

In the next instance, the algorithm is applying a classifier in order to identify coughs. At this stage it is not important to determine whether the cough is productive or not. For this, either the approach described in [34] or the approach described in [35] can be used. Both methods are based on frequency analysis and exploit the characteristics of the frequency content of the cough sounds in different frequency bands in comparison to other sounds that can occur in a pig house (e.g. screams, grunts, feeding systems, contact with doors, etc.). Table 2 provides an overview of the frequency bands an which of those are used by the two algorithms. It should further be noted that both algorithms have employed fuzzy c-means clustering to reach their results and have identified fixed thresholds for their classifiers.

Frequency range (Hz)	Reference [34]	Reference [35]
100-10000	√	√
100-6000	√	√
6000-10000	√	
100-2000	√	
2000-5000	√	√
2000-4000	√	√
4000-6000	√	
2300-3200	√	√
7000-9000	√	

Table 2. Frequency ranges used for the classification of cough. Differences between [34] and [35].

Once a sound has been identified as a cough, then the distinction between a productive and a non-productive cough has to be made. For this, the algorithms presented in [36] or [37] can be used. Both algorithms are based on time-domain characteristics of the coughs. The algorithm presented in [37] is focusing on the decay of the cough sound. To do so, the cough is described by it energy envelope and subsequently a mathematical model is made for the drop of the amplitude. It was shown that productive coughs have a longer decay (as expressed by the time constant) than dry coughs. This is visualised in Fig. 9. It has been hypothesised that this difference is due to differences in lung plasticity that is changing with the occurrence of a respiratory disease. However, further research is needed to confirm this hypothesis.

An alternative method to classify a cough as productive or not is to use the algorithm described in [36]. In this approach, a third order Auto-Regression (AR) model is used (e.g. [38]) to estimate the sound signal. This model has the following form:

$$y_k = a_1 y_{k-1} + a_2 y_{k-2} + a_3 y_{k-3} \tag{1}$$

Where y_k is the value of the signal at sample time k and a_1, a_2 and a_3 are the 3 model parameters. It was shown that productive coughs can be distinguished from dry coughs by using these three parameters. More specifically, a polyhedron is defined in the (a_1, a_2, a_3) space that is subsequently used as a classifier for productive coughs. Alternatively, the centre of the dry cough can be defined and it can be assumed the coughs outside this cluster are productive coughs.

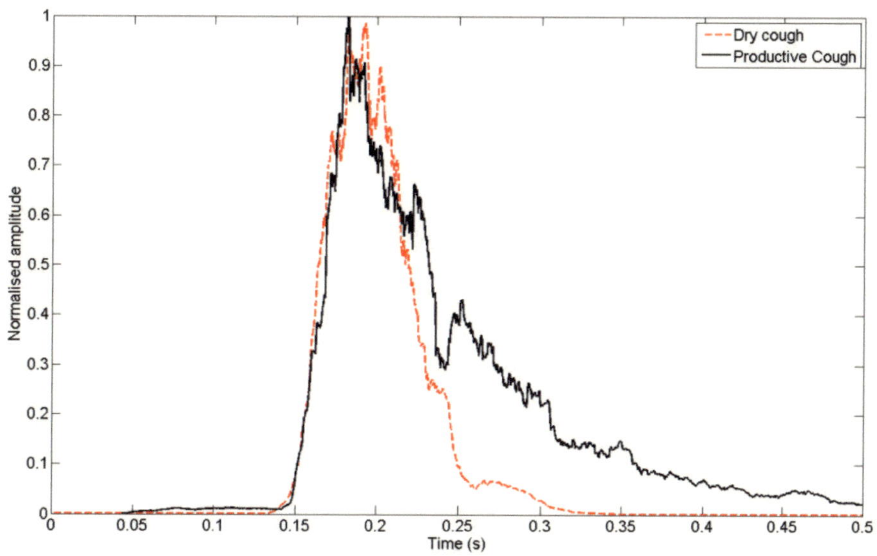

Figure 9. The difference in the decay between a dry (red dashed line) and a productive (black solid line) cough is shown as has been studied in [37]

In the literature it was shown [39] that the vocal tract can act as a first order filter that has the form:

$$y_k = \frac{b_1}{1 + a_1 z^{-1}} u_k \tag{2}$$

Where z^{-1} is the time-shift operator (i.e. $y_k z^{-1} = y_{k-1}$) and u_k is the input to the system. Similarly, (1) can be re-written to have the form:

$$y_k = \frac{1}{1 + a_1 z^{-1} + a_2 z^{-2} + a_3 z^{-3}} e_k \tag{3}$$

Where e_k is white noise.

Since coughing can be considered involuntary in the studied cases, white noise can be considered the input to the system. From the above it can be hypothesised that the classifier was able to identify and exploit some higher order dynamics of the effect of the vocal tract on the produced sound response. Of course this claim has to be studied further.

During the calibration phase of the algorithm, the centre of the dry cough cluster has to be defined. For this, the first five dry coughs need to be identified and the average of their parameters should be used as the centre of the productive cough cluster (the edges of the cluster are defined by the variance of the parameters). It can be assumed that at the beginning of the growing period there is no respiratory disease present and therefore only dry coughs are present. So, the first five coughs that will be introduced are dry coughs and can be used to define the dry cough cluster.

4.2. Validation results

The algorithm has been applied on a continuous recording with a total of 671 sounds (291 productive coughs, 231 dry coughs and 149 other sounds) collected under laboratory conditions.

The following Table 3 shows the results of the classification power of the algorithm.

Sound	Total sounds	True positive identifications	True negative identifications	False positive identifications	False negative identifications
Dry cough	231	-	200	31	-
Productive cough	281	231	-	-	50
Other	144	-	130	14	-

Table 3. Performance of the productive vs. Dry cough identification algorithm

The above algorithm has been integrated in a commercial system by SoundTalks nv. (Belgium – www.soundtalks.be).

http://dx.doi.org/10.5772/56040

5. Conclusion

This chapter has presented three cases where sound has been used to identify and interpret animal vocalisations. The studied cases are focused on identifying the status of animals in livestock production that will allow the caretaker of the animals to quickly identify health, welfare and production issues and take action. On the positive side, sound can be acquired from a large space and without any contact with the animal. This latter is an important factor in the potential for such technology to be applied in practice. Issues such as cost for placement, loss of sensors due to misuse by the animals and acquisition of sensor at the slaughter line are immediately solved. On the negative side though, the acquired sound signal cannot immediately be linked to a specific animal and it is the combination of all sounds and vocalisations in the area. As such, conclusions about a group of animals can be reached but not necessarily for individual ones.

Sound analysis is an excellent show case for the merits of Precision Livestock Farming (PLF) and can already demonstrate the benefit that PLF can have on intensive livestock production. Further research in the field should focus on identifying critical areas for livestock productions and work towards providing automatic tools for early diagnosis. In order for practical implementation, issues that are not related to animal vocalisations will also have to be dealt with. These include, but are not limited to room acoustics, positioning of the microphones for better sound acquisition, identification of vocalisations in a noisy environment, etc.

In the direction of implementation of PLF systems, an EU project (EU-PLF, project number 311825) is currently running with the objective of creating a Blueprint that describes the process of starting from a scientific idea, development of a commercial prototype, implementing a system on farm and creating value for the farmer. In this framework, the results that have been presented in section 4 of this chapter will be applied in a practical setting.

Acknowledgements

We would like to thank the European Union, the IWT (agentschap voor Innovatie door Wetenschap en Technologie), the KU Leuven and Petersime nv. for funding parts of the research described in this chapter. We would also like to thank all the researchers and technicians involved in the research throughout the years.

Author details

Vasileios Exadaktylos, Mitchell Silva and Daniel Berckmans

Division M-BIORES: Measure, Model & Manage Bioresponses, Department of Biosystems, KU Leuven, Belgium

References

[1] Berckmans, D. Precision Livestock Farming- Preface. Computers and Electronics in Agriculture (2008).

[2] Frost, A. R. An overview of integrated management systems for sustainable livestock production. In: Wathes C.M., Frost A.R., Gordon F., Wood J.D. (eds.) Edinburgh, British Society of Animal Science. Occasional Publication (2001). (28), 45-50.

[3] Frost, A, Schofield, C, Beaulah, S, Mottram, T, Lines, J, & Wathes, C. A review of livestock monitoring and the need for integrated systems. Computers and Electronics in Agriculture (1997). , 17(2), 139-159.

[4] Berckmans, D. Automatic on-line monitoring of animals by precision livestock farming. In: ISAH conference "Animal production in Europe : The way forward in a changing world, October 2004, Saint-Malo, France, (2004). , 1, 27-31.

[5] Brehme, U, Stollberg, E, Holz, R, & Schleusener, T. Safer oestrus detection with sensor aided ALT-pedometer. In: Book of Abstracts of the Third International Workshop on Smart Sensors in Livestock Monitoring, 10-11 September 2004, Leuven, Belgium, (2004). , 43-46.

[6] Vranken, E, Chedad, A, Aerts, J-M, & Berckmans, D. Estimation of average body weight of broilers using automatic weighing in combination with image analysis. In: book of abstracts of the Third International Workshop on Smart Sensors in Livestock Monitoring, 10-11 September (2004). Leuven, Belgium, 2004, , 68-70.

[7] Mitchell, K. D, Stookey, J. M, Laturnas, D. K, Watts, J. M, Haley, D. B, & Huyde, T. The effects of blindfolding on behavior and heart rate in beef cattle during restraint. Applied Animal Behaviour Science (2004).

[8] Kohler, S. D, & Kaufmann, O. Quarter-related measurements of milking and milk parameters in an AMS-herd. Milk Science International (2003). , 58, 3-6.

[9] Tona, K. Effects of age of broiler breeders, incubation egg storage duration and turning during incubation on embryo physiology and broiler production parameters. PhD Thesis. KU Leuven, (2003).

[10] Nickelmann, M. Importance of prenatal temperature experience on development of the thermoregulatory constrol system in birds. Thermochimica Acta (2004).

[11] Moriya, K, Pearson, J. T, Burggren, W. W, Ar, A, & Tazawa, H. Continuous measurements of instantaneous heart rate and its fluctuations before and after hatching in chickens. Journal of Experimental Biology (2000). , 203(5), 895-903.

[12] Kemps, B, De Ketelaere, B, Bamelis, F, Decuypere, E, & De Baerdemaeker, J. Vibration analysis on incubating eggs and its relation to embryonic development. Biotechnology Progress (2003). , 19(3), 1022-1025.

[13] Bain, M, Fagan, A, Mullin, M, Mcnaught, I, Mclean, J, & Condon, B. Noninvasive monitoring of chick development in ovo using 7T MRI system from day 12 of incubation through to hatching. Journal of Magnetic Resonance Imaging (2007). , 26(1), 198-201.

[14] Bamelis, F, Kemps, B, Mertens, K, De Ketelaere, B, Decuypere, E, & Debaerdemaeker, J. An automatic monitoring of the hatching process based on the noise of the hatching chicks. Poultry Science (2005). , 84, 1101-1107.

[15] Van Brecht, A, Aerts, J. M, Degraeve, P, & Berckmans, D. Quantification and control of the spatiotemporal gradients of air speed and air temperature in an incubator. Poultry Science (2003). , 82, 1677-1687.

[16] De Smit, L, Bruggeman, V, Tona, J. K, Debonne, M, Onagbesan, O, Arckens, L, De Baerdemaeker, J, & Decuypere, E. Embryonic developmental plasticity of the chick: Increased CO_2 during early stages of incubation changes the developmental trajectories during prenatal and postnatal growth. Comparative Biochemistry and Physiology a-Molecular & Integrative Physiology (2006). , 145, 166-175.

[17] Mortola, J. P. Metabolic response to cooling temperatures in chicken embryos and hatchlings after cold incubation. Comparative Biochemistry and Physiology a-Molecular & Integrative Physiology (2006). , 145, 441-448.

[18] Exadaktylos, V, Silva, M, & Berckmans, D. Real-time analysis of chicken embryo sounds to monitor different incubation stages. Computers and Electronics in Agriculture (2011). , 75, 321-326.

[19] Marx, G, Leppet, J, & Ellendorff, F. Vocalisation in chicks (Galus galus dom.) during stepwise social isolation. Applied Animal Behaviour Science (2001). , 75(1), 61-74.

[20] Ghesquire, K, Van Hirtum, A, Buyse, J, & Berckmans, D. Non-invasive quantification of stress in the laying hen: a vocalization analysis approach. Mededelingen Faculteit Landbouwkundige en Toegepaste Biologische Wetenshcapen Universiteit Gent (2002). , 67(4), 55-58.

[21] Pereira, E. M, Naas, I. A, & Jacob, F. C. Using vocalization pattern to assess broiler's well-being. In: Lokhorst K., Berckmans D. (eds.) Precision Livestock Farming'July 2011, Prague, Czech Republic; (2011). , 11, 11-14.

[22] Bugden, S. C, & Evans, R. M. The development of vocal thermoregulatory response to temperature in embryos of the domestic chicken. Wilson Bulletin (1999). , 111(2), 188-194.

[23] Bugden, S. C, & Evans, R. M. Vocal solicitation of heat as an integral component of the developing thermoregulatory system in young domestic chickens. Canadian journal of Zoology (1997). , 75(12), 1949-1954.

[24] De Moura, J, Naas, I, Alves, E, Crvalho, T, Vale, M, & Lima, K. Noise analysis to evaluate chick thermal comfort. Scientia Agricola (2008).

[25] Feltenstein, M, Ford, N, Freeman, K, & Sufka, K. Dissociation of stress behaviors in the chick social-separation-stress procedure. Physiology & behavior (2002). , 75(5), 675-679.

[26] Mair, J, Marx, G, Petersen, J, & Mennicken, L. Development of multi parametric sound analysis parameters as welfare indicator in chicken flocks. Proceedings of the international seminar on modal analysis (2001). , 3, 1523-1528.

[27] Exadaktylos, V, Berckmans, D, Aerts, J, & Non-invasive, M. Methods for Monitoring Individual Bioresponses in Relation to Health Management. In: Smigorski K (ed.) health Management- Different Approaches and Solutions. Rijeka: InTech; (2011). , 143-160.

[28] Korpáš, I, Sadlonová, J, & Vrabec, M. Analysis of the cough sound: an overview. Puylmomnary Pharmacology (1996). , 9(56), 261-368.

[29] Marx, G, Horn, T, Thielebein, J, Knubel, B, & Von Borell, E. Analysis of pain-related vocalization in young pigs. Journal of Sound and Vibration (2003). , 266(3), 678-698.

[30] Manteuffel, G, Puppe, B, & Schön, P. C. Vocalization of farm animals as a measure of welfare. Applied Animal Behaviour Science (2004).

[31] Schön, P, Puppe, C, & Manteuffel, B. G. Automated recording of stress vocalisation as a tool to document impaired welfare n pigs. Animal Welfare (2004). , 13(2), 105-110.

[32] Schön, P, Puppe, C, & Manteuffel, B. G. Linear prediction coding analysis and self-organizing feature map as tools to classify stress calls of domestic pigs (Sus scrofa). Journal of the Acoustical Society of America (2001). , 110, 1425-1431.

[33] Marx, G, Horn, T, Thielebein, J, Knubel, B, & Von Borell, E. Analysis of pain-related vocalisation in young pigs, Journal of Sound and Vibration (2003). , 266, 687-698.

[34] Van Hirtum, A, & Berckmans, D. Fuzzy approach for improved recognition of citric acid induced piglet coughing from continuous registration. Journal of Sound and Vibration (2003). , 266(3), 677-686.

[35] Exadaktylos, V, Silva, M, Aerts, J, Taylor, M, Berckmans, C. J, & Real-time, D. recognition of sick ppig cough sounds. Computers and Electronics in Agriculture (2008). , 63, 207-214.

[36] Exadaktylos, V, Silva, M, Ferrari, S, Guarino, M, Taylor, C. J, Aerts, J, Taylor, M, Berckmans, C. J, & Time-series, D. analysis for online recognition and localization of sick pig (Sus scrofa) cough sounds. Journal of the Acoustical Society of America (2008). , 124(6), 3803-3809.

[37] Silva, M, Exadaktylos, V, Ferrari, S, Guarino, M, Aerts, J, & Berckmans, M. D. The influence of respiratory disease on the energy envelope dynamics of pig cough sounds. Computers and Electronics in Agriculture (2009). , 69(1), 80-85.

[38] Young, P. C. Recursive Estimation oand Time-Series Analysis- An Introduction for the Student and Practitioner. Berlin: Springer; (2011).

[39] Hannon, B, & Mathias, R. Modelling Dynamic Biological Systems. New York: Springer, (1997).

Chapter 5

Clusterized Mel Filter Cepstral Coefficients and Support Vector Machines for Bird Song Identification

Olivier Dufour, Thierry Artieres, Hervé Glotin and
Pascale Giraudet

Additional information is available at the end of the chapter

http://dx.doi.org/10.5772/56872

1. Introduction

We present here our contribution to the "Machine Learning for Bioacoustics" workshop technical challenge of 30th International Conference on Machine Learning (ICML 2013). The aim is to build a classifier able to recognize bird species one can hear from a recording in the wild.

The method we present here is a rather simple strategy for bird songs and calls classification. It builds on known and efficient technologies and ideas and must be considered as a baseline on this challenge. As we are also co-organizing this challenge, our participation aimed at defining a baseline system, with raw features, that all other participants could compare too. We did not look for optimizing each parameter of our system, and as any other participant, we conducted all the modeling and experimentation applying strictly the rules of the challenge. The method we present is dedicated to the particular setting of the challenge. It relies in particular on the fact that training signals are monolabel, i.e. only one species may be heard, while test signals are multilabeled.

2. Description of the method

We present now the main steps of our approach. The Figures 1 and 2 illustrates the main steps of the preprocessing and of feature extraction.

We consider we want to learn a multilabel classifier from a set of N monolabeled training samples $\{(x^i, y^i) \mid i = 1.. N\}$ where each input x^i is an audio recording and each y^i is a bird species $\forall\ i, y^i \in \{ b_u \mid u = 1.. K\}$ (in our case there are 35 species, K=35). The system should be able to infer the eventually multiple classes (presence of bird species) in a test recording x.

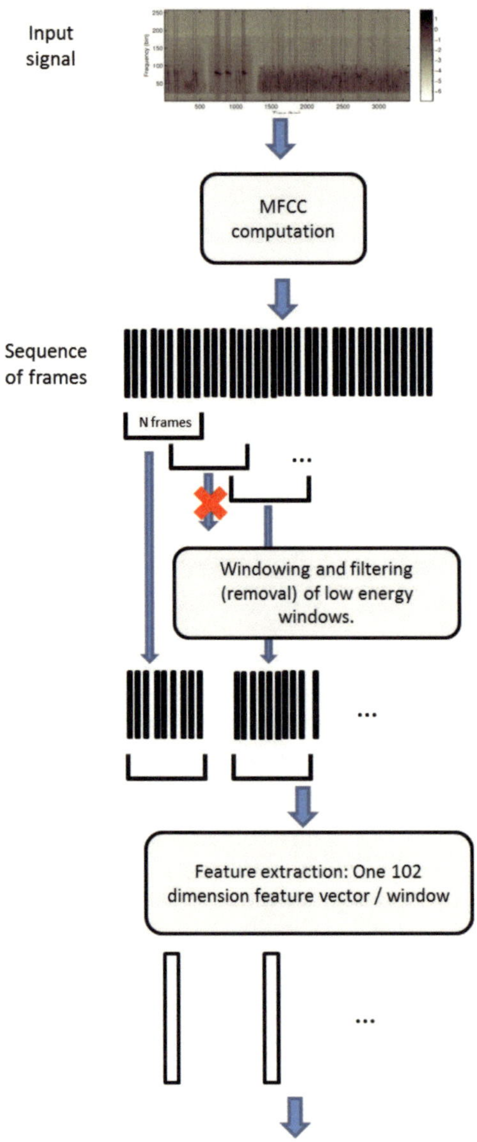

Figure 1. Main steps of the preprocesing and of feature extraction.

2.1. Preprocessing

Our preprocessing is based on MFCC cepstral coefficients, which have been proved useful for speech recognition [4, 11]. A signal is first transformed into a series of frames where each frame consists in 17 MFCC (mel-frequency cepstral coefficients) feature vectors, including energy. Each frame represents a short duration (e.g. 512 samples of a signal sampled at 44.1 kHz).

2.2. Windowing, silence removal and feature extraction

2.2.1. *Windowing*

We use windowing, i.e. computing a new feature vector on a window of n frames, to get new feature vectors that are representative of longer segments. The idea is close to the standard syllable extraction step that is used in most of methods for bird identification [12, 2, 1], but is much simpler to implement. In our case we considered segments of about 0.5 second duration (i.e. n ~ few hundreds of frames) and used a sliding window with overlap (about 80%).

2.2.2. *Silence removal*

We first want to remove segments (windows) corresponding to silence since these would perturbate the training and test steps. This is performed with a clustering step (learnt on training signals) that only considers the average energy of the frames in a window. Ideally this clustering makes that the windows are clustered into silence segments on the one hand, and calls and songs segments on the other hand. Each window with low average energy is considered a silence window and removed from consideration. Our best results were achieved when performing a clustering in three clusters and removing all windows in the lowest energy cluster.

2.2.3. *Feature extraction*

The final step of the preprocessing consists in computing a reduced set of features for any remaining segment / window. Recall that each segment consists in a series of n 17-dimensional feature vectors (with n in the order of hundreds). Our feature extraction consists in computing 6 values for representing the series of n values for each of the 17 MFCC features. Let consider a particular MFCC feature v, let note $(v_i)_{i=1..n}$, the n values taken by this feature in the n frames of a window and let note v_i the mean value of v_i. Moreover let note d and D the velocity and the acceleration of v, which are approximated all along the sequences with $d_i = v_{i+1} - v_i$, and $D_i = d_{i+1} - d_i$. The six features we compute are defined as:

$$f_1 = \frac{\sum_{i=1}^{n}(|vi|)}{n} \quad (1)$$

$$f_2 = \sqrt{\frac{1}{n-1}\sum_{i=1}^{n}\left(v - \bar{v_i}\right)^2} \quad (2)$$

$$f_3 = \sqrt{\frac{1}{n-2}\sum_{i-1}^{n}\left(d-\bar{d}_i\right)^2} \tag{3}$$

$$f_4 = \sqrt{\frac{1}{n-3}\sum_{i-1}^{n}\left(D-\bar{D}_i\right)^2} \tag{4}$$

$$f_5 = \frac{\sum_{i=1}^{n-1}|d_i|}{n-1} \tag{5}$$

$$f_6 = \frac{\sum_{i=1}^{n-2}|D_i|}{n-2} \tag{6}$$

At the end a segment in a window is represented as the concatenation of the 6 above features for the 17 cepstral coefficients. It is then a new feature vector S_t (with *t* the number of the window) of dimension *102*.

Each signal is finally represented as a sequence of feature vectors S_t, each representing duration of about 0.5 second with 80% overlap.

2.3. Training

Based on the feature extraction step we described above the simplest strategy to train a classifier (e.g. we used Support Vector Machines) on the feature vectors S_t which are long enough to include a syllable or a call, with the idea of aggregating all the results found on the windows of a test signal to decide which species are present (see section *Inference* below).

Yet we found that a better strategy was to first perform a clustering in order to split all samples (i.e. S_t) corresponding to a species into two different classes. The rationale behind this process is that calls and songs of a particular species are completely different sounds [9] so that corresponding feature vectors S_t probably lie in different areas in the feature space. It is then probably worth using this prior to design classifiers (hopefully linear) with two times the number of species rather than using non linear classifiers with as many classes as there are species.

We implemented this idea by clustering all the frames S_t for a given species into two or more clusters. The two clusters are now considered as two classes that correspond to a single species. At the end, a problem of recognizing *K* species in a signal turns into a classification problem with 2 x *K* classes. Note also that since the setting of the challenge is such that there is only one species per training signal, all feature vectors S_t of all signal of a given bird species b_u that fall into cluster one are labeled as belonging to class b_u^1 and all that fall into cluster 2 are labeled as belonging to class b_u^2.

The final step is to learn a multiclass classifier (SVM) in a one-versus-all fashion, i.e. learning one SVM to classify between the samples from one class and the samples from all other classes. This is a standard approach (named Binary Relevance) for dealing with multilabel classification problem where one sample may belong to multiple classes. It is the optimal method with respect to the Hamming Loss, i.e. the number of class prediction errors (either false positive and false negative).

2.4. Inference

At test time an incoming signal is first preprocessed as explained before in section 2.1, silence windows are removed (using clusters), and feature extraction is performed for all remaining segments. This yields that an input signal is represented as a series of m feature vectors S_t.

All these feature vectors are processed by all $2K$ binary SVMs which provide scores that are interpreted as class posterior probabilities (we use a probabilistic version of SVM), we then get a matrix $m \times 2K$ of scores P (c | S_t) with $c \in \{b_u^j \mid u=1..\,K,\ j=1,2\}$ and t = 1..m.

We experimented few ways to aggregate all these scores into a set of K scores, one for each species, enabling ranking the species by decreasing probability of occurrence. Indeed this is the expected format of a challenge submission, from which an AUC (Area Under the Curve) score is computed. First we compute $2K$ scores, one for each class, then we aggregate the scores of the two classes of a given species.

Our best results were obtained by computing mean probabilities of all scores { P (c | s_t) | t=1..m } for each class c, using harmonic mean or trimmed harmonic mean (where a percentage of the lowest scores are discarded before computing the mean). This yields scores that we consider as class posterior probabilities of classes given the input signal x, P (c | x).

The ultimate step consists in computing a score for each species b_u given the scores of the two corresponding classes b_u^1 and b_u^2. We used the following aggregation formulae:

$$P(b_u \mid x)=1-\left(1-P\left(b_u^1 \mid x\right)\right) \times \left(1-P\left(b_u^2 \mid x\right)\right) \tag{7}$$

3. Experiments

3.1. Dataset

We describe now the data used for the "Machine Learning for Bioacoustics" technical challenge. Note that the training dataset (signals with corresponding ground truth) was available for learning systems all along the challenge together with the test set, without ground truth. Participants were able to design their methods and select their best models by submitting predictions on the test set which were scores on a subset only of the test set (33%). The final evaluation and the ranking of participants were performed on the full test set once all participants have selected 5 of all their systems submitted.

Training data consisted in thirty-five 30-seconds audio recordings labeled with a single species; there was one recording per species (35 species overall). Yet, some train recording can include low signal-to-noise ratio (SNR) signals of a second bird species of bird. Moreover, according to circadian rhythm of each species, other acoustically actives species of animals can be present such as nocturnal and diurnal insects (Grylidae, Cicada).

Test data consisted in ninety 150-seconds audio recordings with possibly none or multiple species occurring in each signal.

The training and test data recordings have been performed with various devices in various geographical and climatological settings. In particular background and SNR are very different between training and test. All wav audio recordings have been sampled at 44 100 Hz with a 16-bits quantification resolution. Recordings were performed with 3 Song Meter SM2+ (Wildlife Acoustic recording device). Each SM2+ has been installed in a different sector (A, B and C) of a Regional Park of the Upper Chevreuse Valley.

Every SM2+ recorded, at the same dates and hours (between 24 03 2009 and 22 05 2009), one 150-seconds recording per day between 04h48m00s a.m. and 06h31m00s a.m., which correspond to the maximal acoustical bird-activity period.

3.2. Implementation details

3.2.1. Frames and overlapping sizes

We computed Mel-frequency cepstral coefficients (MFCC) with the *melfcc.m Matlab* function from ROSA laboratory of Columbia University [8]. This function proposes 17 different input parameters. We tested numerous possible configurations [7] and measured for each one the difference of energy contained in a given train file and a reconstructed signal of this recording based on cepstral coefficients.

The difference was minimal with following parameters values:

window=512, fbtype=mel, broaden=0, maxfreq=sr/2, minfreq=0,wintime=window/sr, hoptime=wintime/3, numcep=16,usecmp=0, dcttype=3, nbands=32, dither=0, lifterexp=0,sumpower=1, preemph=0, modelorder=0, bwidth=1, useenergy=1

This process transforms a 30-seconds train audio recording (at 44 kHz sampling rate) into about 7 700 frames of 16 cepstral coefficients which we augmented with the energy computed by setting *useenergy=0*.

Next we computed feature vector S_t on 0.5 second windows with 80% overlap, which yields about *n=300* feature vectors per training signal (hence per species since there is only one training recording per species) and about *m=* feature vectors per test signal.

3.2.2. LIBSVM settings

We used a multiclass SVM algorithm based on LIBSVM [3]. We selected model parameters (kernel type etc.) through two fold cross validation. Best scores have been obtained with C-SVC SVM type and linear kernel function.

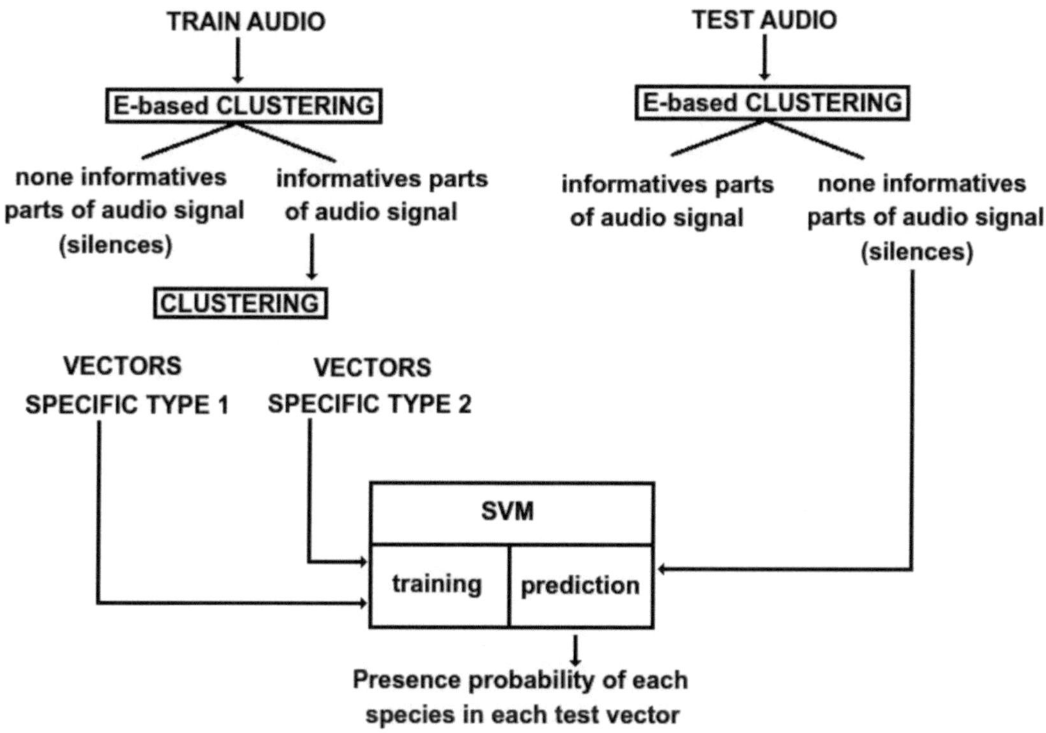

Figure 2. Technical principle of our best-scored run

4. Results

4.1. General results

We report only our best results that correspond to the method presented in this paper for various computations for the class score at inference time.

Table 1 shows how the way the mean score of a class is computed on the test set (see section 2.4) and influences the final result. The table compares arithmetic mean, harmonic mean, and trimmed arithmetic mean (at 10, 20 et 30%). A trimmed mean at p% is the arithmetic mean computed after discarding p% extreme values, i.e. the *p*/2% lowest values and the *p*/2% largest values.

Although our method is simple it reached the fourth rank over more than 77 participating teams at the Kaggle ICML Bird challenge with a score of *0.64639* while the best score (Private score) of all challengers was 0.694 (the corresponding public Leaderboard score was 0.743). See [13] for the best system, and [14] for the description of the other systems. It is also worth noting that our system ranked about fifteen only on the validation set (one third of the total test set). This probably shows that our system being maybe simpler than other methods exhibits at the end a more robust behavior and improved generalization ability.

mean aggregation	Private score	Public score
arithmetic mean	0.61362	0.63974
harmonic mean	0.64234	0.67344
trimmed mean 10%	0.64158	0.68612
trimmed mean 20%	0.64639	0.69163
trimmed mean 24%	0.64699	0.69103
trimmed mean 30%	0.64614	0.68881

Table 1. Score Kaggle icml (AUC) according to the way scores are aggregated. Public scores are calculated on a third of the test data, while private scores are calculated on the other part. Only the private scores are the official competition results. The best private score of all challengers is 0.694 [13].

4.2. Monospecific results

According to these scores for 7 species, we notice in Figure 3:

- Scores of our model are close to the best ones and evolve the same way for the concerned species. The slight difference is probably due to the way we calculate (trimmed mean) the presence probability of one given species in a 150-seconds recording compared to the presence probability of this same species in a half-second frame.
- All teams were not able to score high for *Columba palumbus* (Common Wood-pigeon), *Erithacus rubecula* (European Robin), *Parus caeruleus* (Blue Tit), *Parus palustris* (Marsh Tit), *Pavo cristatus* (Blue Peafowl) and *Turdus viscivorus* (Mistle Thrush).
    - In the Common Wood-pigeon (top of Figure 4) train recording, we can see a series of 5 syllables (around 500 Hz). Syllables are very stable and different. Their alternation in time domain is strict. Also, the train recording is highly corrupted by cicadas between 4 and 6 kHz and in the test recording, SNR is low. The series last 2.5 seconds (compared to 4 seconds in TRAIN) and are composed of 6 syllables well differentiated.
    - The European Robin (bottom of Figure 4) is typically bird species whose songs are diverse and rich in syllables. Frequency-domain variability between different songs and syllables is important. Song duration varies between 1.5 and 3 seconds. It is one of the rare species that can emit up to 8 kHz.
    - In Blue Tit train recording, other species of birds are present. Therefore, Blue Tit produces 5 different cries composed of 5 different syllables.
    - Mistle Thrush train recording songs vary a lot and are very different from songs in the test recordings.

MFCC compression has the property of lowering the weights of cepstral coefficients corresponding to higher frequencies of the spectrogram. As a result, MFCC can lead to losing a part of the signal that may be important in European Robin's case. Futhermore, the high variability of the cries or songs of the different species is difficult to manage by classifiers, especially when they are constrained to retain and learn only 2 types of emissions per species. Considering two types of emissions was particularly sub-optimal for these 3 cases.

http://dx.doi.org/10.5772/56872

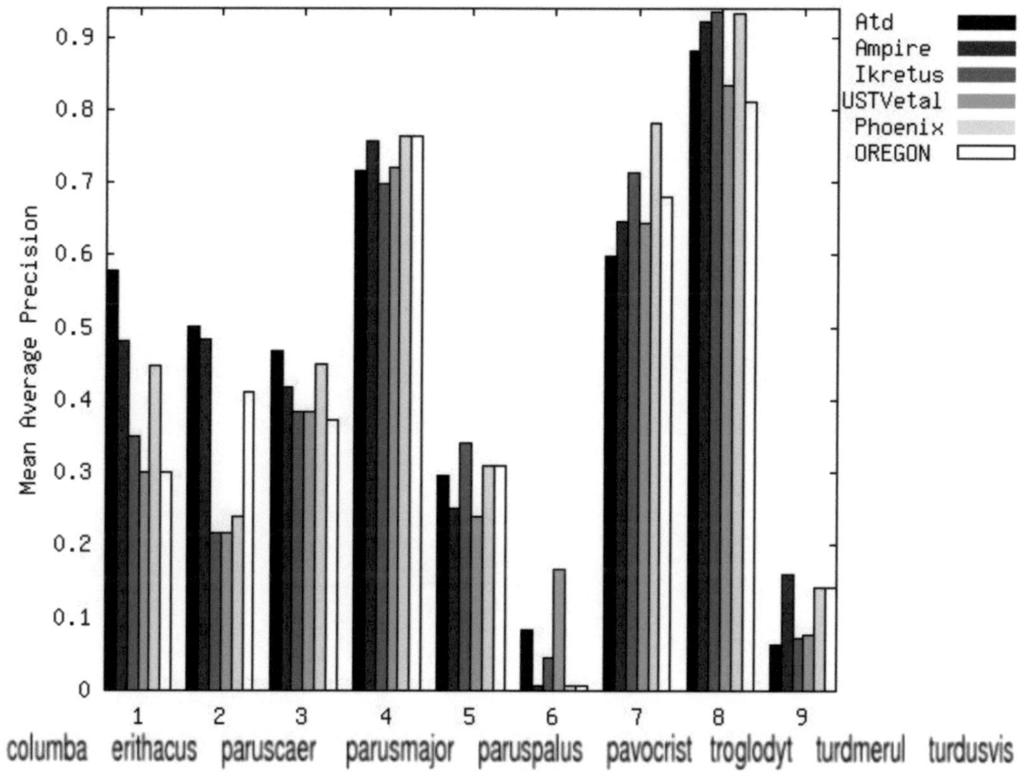

Figure 3. Mean Average Precision (MAP) scores on nine species (ordered in abscissa from left to right) of the 6 best teams of the challenge. The label 'USTVetal' refers to our team (MAP was not the official metrics of the challenge but give interesting comparisons).

- For all teams, scores were very satisfactory for *Parus major* (Great Tit), *Troglodytes troglodytes* (Winter Wren) and *Turdus merula* (Eurasian Blackbird).
  - Great Tit's signals (middle of Figure 4) are very simple and periodically repeated. A 500-hertz high-pass filter has been applied on the train recording.
  - Winter Wren's acoustic patterns are really stable. A 1000-hertz high-pass filter has been applied on the train recording.
  - Eurasian Blackbird's train recording has been filtered by a band pass filter from 1-6 kHz. Best Mean Average Precisions were obtained when low frequency and high-frequency noise was removed by filtering.

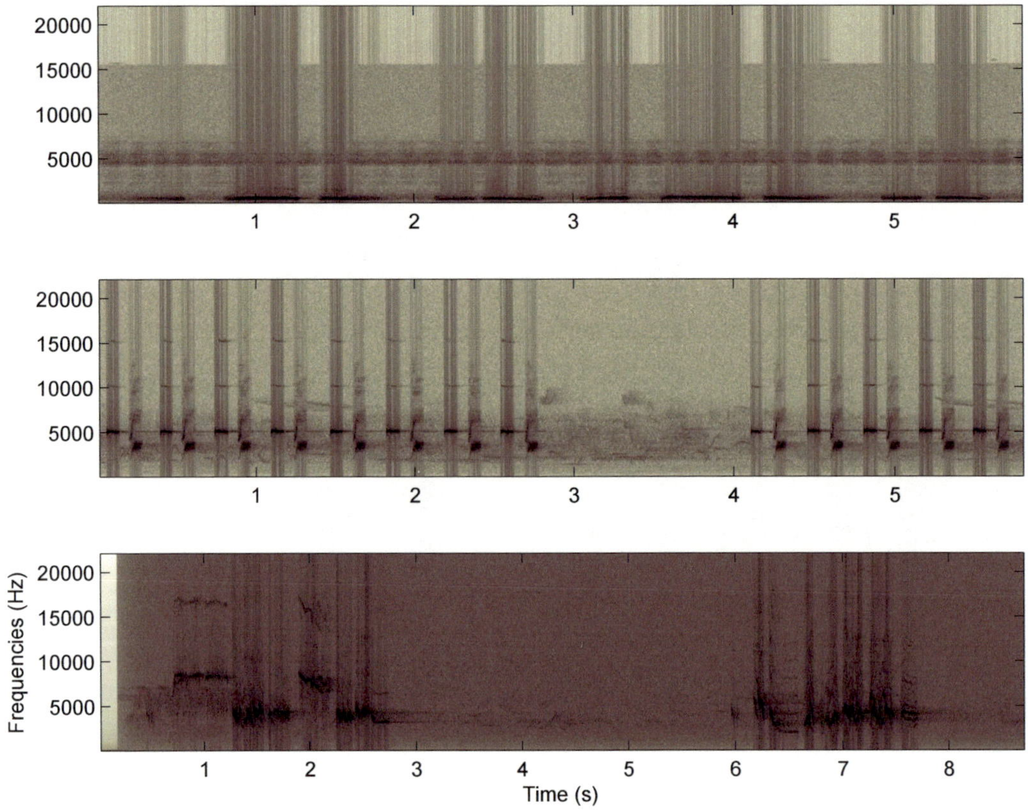

Figure 4. Time-Frequential spectrograms of train recording's extracts. From top to bottom: Common Wood-pigeon, Great Tit and European Robin.

We assume that congruence observed between the scores of the 6 best teams for these 9 species is the same considering each of the species. The fact that the scores of each species evolve the same way indicates that the Mean Average Precision (M.A.P) differences between species can be due to:

1. Some species produce sounds harder to characterize than others: strong variability in frequency and/or temporal domain.
2. Train recordings can't be compared to test recordings regarding SNR: filters, harmonic richness, source-microphone distances etc. differ a lot.
3. Signals of interest are easier to extract in some train recordings than in others because of data acquisition. Some filters have been applied to a part of the train recordings.

4. For a given species, the signals provided in the train recording may not include a global repertoire and this way not be part of the respective species test recordings.

5. For each species, frequency content of emissions and location of source in its environment differ widely. Each bird species uses the available space in an ecosystem differently. Obstacles between source and microphone depend on diet and customs of species (arboricol, walking, granivorous, insectivorous species etc). But all frequencies aren't affected the same way by transmission loss in the environment. For example, low frequencies are particularly well filtered by vegetation close from the ground. Common Wood-pigeon typically emits in low frequencies (see figure 4).

6. Natural (rain, wind, insects) or anthropic (motors etc) acoustic events are more diverse and strong (regarding energy) in test recordings than in train. In addition, these events vary much from one species to an other.

Hence, it seems reasonable to affirm that more complex syllables extraction methods (segmentation step) combined with the MFCC way constitute a better solution to improve our performance. They would allow us to retain intraspecific variability for each class and eliminate non-relevant information.

5. Conclusion and perspectives

Although the method that we presented is simple it performed well on the challenge and was much robust between validation step and test set. We believe this robustness comes from the simplicity of the method that do not rely on complex processing steps (like identifying syllables) that other participants could have used [10, 13, 15, 16].

Possible improvements would consist in the integration in the model of additional information such as syllables extraction, weather condition, or a taxonomia of species, allowing more accurate hierarchical classification schemes. Also the MFCC shall be replaced either by a scattering transform [17] or a deep convolutional network [18], that build invariant, stable and informative signal representations for classification.

Acknowledgements

We thank Dr. Xanadu Halkias for her useful comments on this paper. This work is supported by the MASTODONS CNRS project Scaled Acoustic Biodiversity SABIOD and the Institut Universitaire de France that supports the "Complex Scene Analysis" project. We thank F. Jiguet and J. Sueur and F. Deroussen [6, 5] who provided the challenge data.

PhD funds of 1st author are provided by Agence De l'Environnement et de la Maîtrise de l'Energie (mila.galiano@ademe.fr) and by BIOTOPE company (Dr Lagrange, hlagrange@biotope.fr, R&D Manager).

Author details

Olivier Dufour[1,2], Thierry Artieres[3], Hervé Glotin[1,2,4] and Pascale Giraudet[1]

*Address all correspondence to: olivierlouis.dufour@gmail.com, thierry.artieres@lip6.fr, glotin@univ-tln.fr, giraudet@univ-tln.fr

1 Université du Sud Toulon Var, France

2 Aix-Marseille Université, CNRS, ENSAM, LSIS, Marseille, France

3 LIP6, Université Paris VI, France

4 Institut Universitaire de France, Paris, France, France

References

[1] F. Briggs, X. Fern, and R. Raich. Acoustic classification of bird species from syllables : an empirical study. Technical report, Oregon State University, 2009.

[2] F. Briggs, B. Lakshminarayanan, L. Neal, X. Fern, R. Raich, M. Betts, S. Frey, and A. Hadley. Acoustic classification of multiple simultaneous bird species : a multi-instance multi-label approach. Journal of the Acoustical Society of America, 2012.

[3] C.-C. Chang. Libsvm. http://www.csie.ntu.edu.tw/~cjlin/libsvm/, 2008.

[4] L. Chang-Hsing, L. Yeuan-Kuen, and H. Ren-Zhuang. Automatic recognition of bird songs using cepstral coefficients. Journal of Information Technology and Applications Vol. 1, pp.17-23, 2006.

[5] F. Deroussen. Oiseaux des jardins de france. Nashvert Production, Charenton, France, 2001. naturophonia.fr.

[6] F. Deroussen and F. Jiguet. Oiseaux de france, les passereaux, 2011.

[7] O. Dufour, H. Glotin, T. Artières, and P. Giraudet. Classification de signaux acoustiques : Recherche des valeurs optimales des 17 paramètres d'entrée de la fonction melfcc. Technical report, Laboratoire Sciences de l'Information et des Systèmes, Université du Sud Toulon Var, 2012.

[8] D. P. W. Ellis. PLP and RASTA (and MFCC, and inversion) in Matlab, 2005. online web resource.

[9] S. Fagerlund. Acoustics and physical models of bird sounds. In Seminar in acoustics, HUT, Laboratory of Acoustics and Audio Signal Processing, 2004.

[10] H. Glotin and J. Sueur. Overview of the first international challenge on bird classification, 2013. online web resource.

[11] A. Michael Noll. Short-time spectrum and cepstrum techniques for vocal-pitch detection. Journal of the Acoustical Society of America, Vol. 36, No. 2, pp. 296-302, 1964.

[12] L. Neal, F. Briggs, R. Raich, and X. Fern. Time-frequency segmentation of bird song in noisy acoustic environments. In International Conference on Acoustics, Speech and Signal Processing, 2011.

[13] Rafael Hernandez Murcia, "Bird identification from continuous audio recordings", in Proc. of first int. wkp of Machine Learning for Bioacoustics ICML4B joint to ICML 2013, Ed. H. Glotin et al, Atlanta, ISBN 979-10-90821-02-6 http://sis.univ-tln.fr/~glotin/ICML4B2013_proceedings.pdf

[14] H. Glotin, Y. Lecun, P. Dugan, C. Clark, X. Halkias, Proceedings of the first int. wkp of Machine Learning for Bioacoustics ICML4B joint to ICML 2013, Ed. H. Glotin et al, Atlanta, ISBN 979-10-90821-02-6 http://sis.univ-tln.fr/~glotin/ICML4B2013_proceedings.pdf

[15] Briggs et al., "ICML 2013 Bird Challenge – Tech Report, in Proc. of first int. wkp of Machine Learning for Bioacoustics ICML4B joint to ICML 2013, Ed. H. Glotin et al, Atlanta, ISBN 979-10-90821-02-6 http://sis.univ-tln.fr/~glotin/ICML4B2013_proceedings.pdf

[16] Dan Stowell and Mark D. Plumbley, " Acoustic detection of multiple birds in environmental audio by Matching Pursuit",in Proc. of first int. wkp of Machine Learning for Bioacoustics ICML4B joint to ICML 2013, Ed. H. Glotin et al, Atlanta, ISBN 979-10-90821-02-6 http://sis.univ-tln.fr/~glotin/ICML4B2013_proceedings.pdf

[17] J. Andén and S. Mallat, "scattering transform applied to audio signals and musical classification" Proceedings of International Symposium on Music Information Retrieval (ISMIR'11), 2011

[18] Mikael Henaff, Kevin Jarrett, Koray Kavukcuoglu and Yann LeCun: Unsupervised Learning of Sparse Features for Scalable Audio Classification, Proceedings of International Symposium on Music Information Retrieval (ISMIR'11), (Best Student Paper Award), 2011

Section 2

Human Hearing Estimations and Cognitive Soundscape Analysis

Chapter 6

Head-Related Transfer Functions and Virtual Auditory Display

Xiao-li Zhong and Bo-sun Xie

Additional information is available at the end of the chapter

http://dx.doi.org/10.5772/56907

1. Introduction

1.1. Sound source localization and HRTFs

In real environments, wave radiated by sound sources propagates to a listener by direct and reflected paths. The scattering, diffraction and reflection effect of the listener's anatomical structures (such as head, torso and pinnae) further disturb the sound field and thereby modify the sound pressures received by the two ears. Human hearing comprehensively utilizes the information encoded in binaural pressures and then forms various spatial auditory experiences, such as sound source localization and subjective perceptions of environmental reflections.

Psychoacoustic experiments have proved that the following cues encoded in the binaural pressures contribute to directional localization [1]:

1. The interaural time difference (ITD), i.e., the arrival time difference between the sound waves at left and right ears, is the dominant directional localization cue for frequencies approximately below 1.5 kHz.

2. The interaural level difference (ILD), i.e., the pressure level difference between left and right ears caused by scattering and diffraction of head etc., is the important directional localization cue for frequencies approximately above 1.5 kHz.

3. The spectral cues encoded in the pressure spectra at ears, which are caused by the scattering, diffraction, and reflection of anatomical structures. In particular, the pinna-caused high-frequency spectral cue above 5 to 6 kHz is crucial to front-back disambiguity and vertical localization.

4. The dynamic cue, i.e., the change in binaural pressures (thus ITD and ILD) introduced by head movement, also contributes significantly to front-back disambiguity and vertical localization.

In this chapter, the sound source position is specified by a spherical coordinate (r, θ, ϕ), where r denotes the source distance relative to the head center (i.e., the origin). Elevation ϕ varies from –90° to 90° with –90°, 0°, 90° denoting below, horizontal and above, respectively. Azimuth θ varies from 0° to 360° with θ = 0°, 90°, 180°, and 270° denoting front, right, behind, and left in the horizontal plane, respectively.

When both sound source and listener are fixed, the acoustical transmission from a point source to the two ears can be regarded as a linear-time-invariable (LTI) process (see Figure 1). Head-related transfer functions (HRTFs) are defined as the acoustical transfer function of this LTI system:

$$H_L(r,\theta,\phi,f,a) = \frac{P_L(r,\theta,\phi,f,a)}{P_0(r,f)}, \quad H_R(r,\theta,\phi,f,a) = \frac{P_R(r,\theta,\phi,f,a)}{P_0(r,f)}. \tag{1}$$

where P_L and P_R represent sound pressures at left and right ears, respectively; P_0 represents the free-field sound pressure at head center with the head absent. Generally, HRTFs vary as functions of frequency f and source position (r, θ, ϕ) (distance and direction) as well as individual a. For $r > 1.0 - 1.2$ m, HRTFs are approximately independent of source distance and called far-field HRTFs. For $r < 1.0$ m, however, HRTFs are relevant to source distance and called near-field HRTFs.

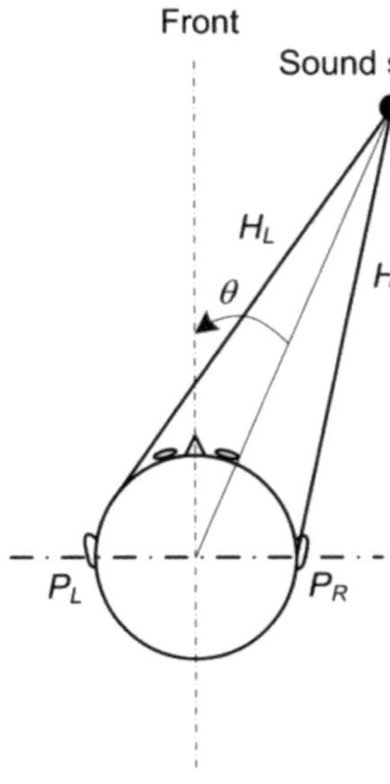

Figure 1. Acoustic transmission from a point sound source to the two ears

The measurement point for P_L and P_R in Eq. (1) varies across studies, among which the eardrum is a most natural choice. Since the external ear canal is proved to be a direction-independent one-dimensional transmission line below 10 kHz, the binaural pressures can be measured at an arbitrary point from the blocked or open entrance of ear canal to the eardrum [2]. Although the pressures differ at different reference points, they all capture the directional information of sound source.

The time-domain counterparts of HRTFs are known as head-related impulse responses (HRIRs), which relate to HRTFs by Fourier transform. HRIRs are the impulse responses from a point sound source to two ears in the free-field. More generally, in reflective environments such as a room, the impulse responses from a source to two ears are called binaural room impulse responses (BRIRs). BRIRs can be regarded as generalized HRIRs from a free-field without reflections to a sound field with reflections.

HRTFs or HRIRs contains most of above-mentioned source localization cues, except the dynamic cue caused by head movement. Therefore, they are vital to the study of binaural localization [3]. One important application of HRTFs is the binaural synthesis in virtual auditory display (VAD). These are the major contents of this chapter.

2. Obtainment of HRTF

2.1. Measurement

Measurement is a conventional and accurate way to obtain HRTFs, especially for human individuals. The principle and methods for HRTF measurement are similar to those for measuring the response of an acoustical LTI system. Figure 2 shows a typical block diagram of HRTF measurement. The measuring signal generated by a computer is rendered to a loudspeaker after passing through a D/A converter and a power amplifier. Resultant signals are recorded by a pair of microphones positioned at subject's two ears, and then delivered to the computer after amplification and A/D conversion. Finally, HRTFs or HRIRs are obtained after some necessary signal processing.

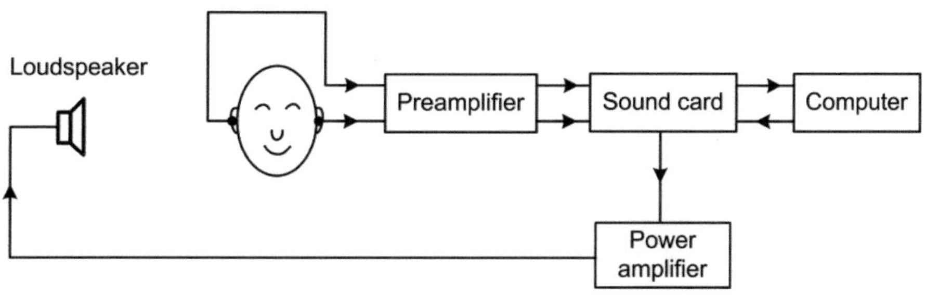

Figure 2. The block diagram of HRTF measurement

To avoid the influence of environment, measurements are usually undertaken in an anechoic chamber, or in a reflective room with a succeeding time-domain windowing so as to rule out reflections. Post-equalization is usually supplemented to correct the non-ideal transmission response in the measurement chain in Figure 2.

Due to the one-dimensional transmission characteristic from the entrance of ear canal to the eardrum, the binaural pressures can be recorded at an arbitrary point from entrance of ear canal to eardrum. In practice, recording binaural pressures with miniature microphones at the blocked ear cannel entrance is the most convenient method for HRTF measurements of human subjects, see Figure 3.

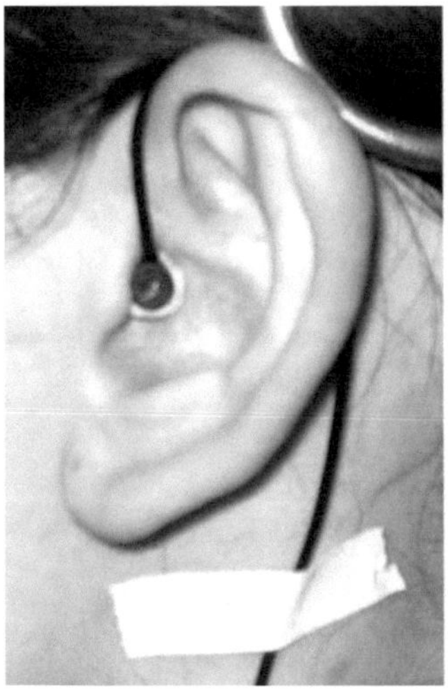

Figure 3. Blocked-ear-canal measurement with miniature microphone

Various signals, such as impulse signals, exponential sweep signals, and pseudo-random noise signals, have been used in HRTF measurement, among which the bipolar maximal length sequence (MLS) is often used. The MLS is a pseudo-random noise sequence (signal) with a deterministic and periodic structure, but possesses characteristics similar to a random noise. In particular, it possesses the lowest crest factor and pulse-like autocorrelation function (equivalent to a nearly uniform power spectrum). For a long N-point MLS, its HRIR $h(n)$ is related to the circle cross-correlation calculation between the recorded signal y and MLS signal x as:

$$R_{xy}(n) \approx h(n) - \frac{1}{N} \sum_{n=0}^{N-1} h(n). \tag{2}$$

One advantage of the measurement using the MLS-like pseudo-random noise sequence is its noise immunity. The deterministic and periodic characteristics of the MLS allow a high signal-to-noise ratio in measurement by means of averaging. In addition, the low cross-correlation among the time-order-reversed MLS also allows for a fast measurement of HRTFs at different directions using multiple sources simultaneously [4].

Figure 4 is the photo of a set of computer-controlled HRTF measurement apparatus in our laboratory [5]. Multiple sound sources (i.e., small loudspeakers) are arranged in different elevations. A computer-controlled horizontal turntable is adopted, on which a rod is installed to support the artificial head or a seat for a human subject. The source distance relative to the head center is adjustable with a maximum distance of 1.2 m.

Thus far, some research groups have constructed databases for measured far-field HRTFs from artificial heads or human subjects [6-13]. Some databases are available on the internet. Foremost of these are the HRTFs of Knowles Electronic Manikin for Acoustic Research (KEMAR), an artificial head-and-torso model for the research of binaural hearing, see Figure 4. The KEMAR HRTF database constructed by the MIT Media laboratory has been widely used in numerous studies. The database contains 512-point far-field ($r = 1.4$ m) HRIRs of 710 spatial directions from elevation −40° to 90°. In the measurements, the binaural pressures were recorded at the ends of the occluded-ear simulator, i.e., at eardrums.

Figure 4. Photo of HRTF measurement apparatus in our lab.

However, the HRTFs of an artificial head merely represent the mean characteristics of a certain population, based on which the artificial head was designed, rather than the individual characteristics of humans. For human HRTFs, the CIPIC database consists of 43 subjects mainly from western population [10]. There are statistically significant differences in anatomical dimensions and shapes as well as resulting HRTFs among different populations. Thus, our group measured and established a far-field HRTF database with 52 Chinese subjects (half males and half females) in 2005 [13]. This database includes far-field 512-point HRTFs at 493 source directions per subject with 44.1 kHz sampling frequency and 16-bit quantization. The database also includes 17 anthropometric parameters relating to dimensions of head and pinna, and so on.

Near-field HRTF measurement is relatively difficult. First, a near-field point sound source is urgently needed. In the case of near-field, an ordinary small-size loudspeaker system is no long approximately being as a point sound source due to its size, directivity, and multiple scattering between source and subject. Second, near-field HRTF measurement is much more time-consuming because measurements at various distances are required due to the distance dependency of near-field HRTF. Such tedious measurement process is particularly unbearable for human subjects. Till now, only a few research groups have measured near-field HRTFs for artificial heads, and no public database is available [14-16]. Based on a spherical dodecahedron sound source, Yu et al. measured the near-field HRTF for KEMAR with DB 60/61 small pinnae [17]. The binaural pressures were recorded at the ends of a pair of Zwislocki occluded-ear simulators. The resultant database includes HRIRs at 10 source distances of 0.20, 0.25, 0.30, 0.40, 0.50, 0.60, 0.70, 0.80, 0.90, and 1.00 m, and 493 directions at each source distance. Each HRIR is 512-point length with 44.1 kHz sampling frequency and 32-bit (float) quantization.

2.2. Computation

Computation is an alternative method for obtaining HRTFs. From mathematical and physical perspectives, calculating HRTFs pertains to solving the scattering problem caused by the human anatomical structures; that is, solving the wave or Helmholtz equation subject to certain boundary conditions.

The analytical solution of HRTFs can be solved from some simplified human anatomical geometry. The spherical-head model is the simplest model for HRTF calculation. As shown in Figure 5, the head is simplified as a rigid sphere with radius a, and the ears as two opposite points on the sphere. For an incident plane wave or a sinusoidal point source that is infinitely distant from the sphere center, the far-field HRTF can be calculated by Rayleigh's solution for pressure at the sphere surface, as [18]

$$P(\Gamma, f) = -\frac{P_0}{(ka)^2} \sum_{l=0}^{\infty} \frac{(2l+1) j^{l+1} P_l(\cos\Gamma)}{dh_l(ka)/d(ka)}, \tag{3}$$

where Γ is the angle between incident direction and received point (ear) on the sphere surface; $k = 2\pi f / c$ is the wave number; $P_l(\cos\Gamma)$ is the Legendre polynomial of degree l; $h_l(ka)$ is the lth-order spherical Hankel function of the second kind. The calculation of spherical-head HRTF can be extended to the case of an arbitrary (finite) source distance [19].

To investigate the torso effect on HRTFs, a simplified head-and-torso model called the snowman model was used for HRTF calculation [20]. The model consists of a spherical head located above a spherical torso, and the HRTFs of the model can be solved using the method of multi-scattering or multipole re-expansion [21].

The calculation from the simplified head-and-torso model reflects some basic features of HRTFs, but it is roughly valid at low and mid frequencies below 3 kHz. The geometry of a real human head is more complex than a sphere and the contribution of pinnae to high-frequency HRTFs is significant. To improve HRTF calculation accuracy, some numerical methods such

http://dx.doi.org/10.5772/56907

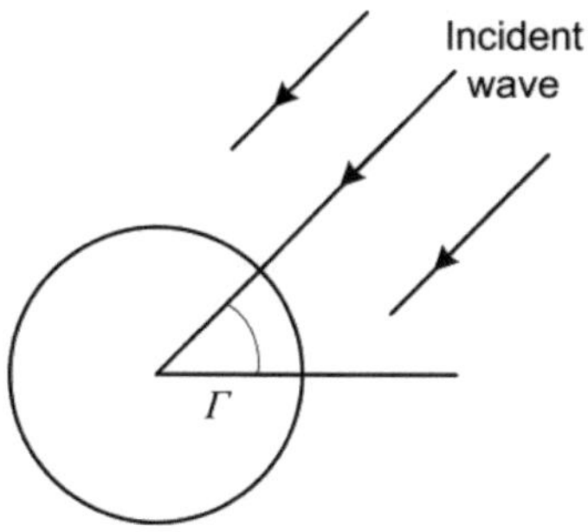

Figure 5. The spherical-head model for HRTF calculation

as boundary element method (BEM) have been developed [22-24]. In BEM calculation, the solution to the scattering problem of human anatomical structures can be expressed as a Kirchhoff–Helmholtz integral equation. The geometrical surfaces of a human or artificial head (such as head and pinnae) are first acquired by a laser 3D scanner or other scanning devices, and then discretized into a mesh of triangular elements. The largest length of the elements is closely related to the maximal frequency to be analyzed and should not exceed 1/4 to 1/6 of the shortest wavelength concerned. Consequently, the Kirchhoff –Helmholtz integral equation is converted into a set of linear algebra equations. Currently, the BEM calculation yields HRTFs with reasonable accuracy up to or near 20 kHz.

However, numerical methods are also time-consuming. It usually takes dozens to hundreds of hours for a typical personal computer to calculate a set of HRTFs at various source directions by conventional BEM (depending on computational power, the number of elements, frequency, and spatial resolution, etc.). High computational costs make calculation difficult. To reduce the computational cost, the acoustic principle of reciprocity can be incorporated in HRTF calculation. According to the acoustic principle of reciprocity, interchanging the source/ receiver positions results in identical pressures. In HRTF calculation, therefore, source position can be fixed at the two ears and receiver points are selected at various spatial directions outside the body. There is still some calculation due to each receiver, but these calculations are much faster than the conventional calculation [23]. Moreover, some researches proposed a fast multipole accelerated boundary element method (FMM BEM) for HRTF calculation [25].

2.3. Customization

Aside from measurement and calculation, in practical use, individualized HRTFs can also be approximately obtained by customization. Generally, HRTFs can be customized using anthropometry-based or subjective-selection-based methods.

The anthropometry-based methods hypothesize that there exists a strong relationship between individual HRTFs and individual anatomical features, because HRTFs characterize the interaction between incident sound waves and human anatomical structures. Accordingly, the individualized HRTFs can be approximately estimated or matched from appropriate anatomical measurements and a baseline database of HRTFs. Practical customization methods include selecting the best-matched HRTFs from a baseline database in terms of the similarity on the measured anatomical parameters among the subject and those in the baseline database [26]; scaling the logarithmic HRTF magnitude from a generic HRTF using anthropometry-predicted scale factor [27]; establishing statistical relationship between the parameterized representation of HRTFs and anatomical parameters, and then predicting the parameters for HRTF representation by anthropometric measurements [28]. The subjective-selection-based methods approximately evaluate the individual HRTFs by appropriate subjective evaluation schemes so as to achieve improved perceived performance, such as localization performance in VAD [29, 30].

Customization of individual HRTFs usually necessitates a baseline database with adequate subjects so as to adapt to the diversity in individualized HRTFs. Customization is simpler than measurement or calculation and yields moderate results, but its accuracy is inferior to measurement and calculation.

3. Physical characters of HRTF

3.1. Time- and frequency-domain characteristics

Although HRIRs or HRTFs vary across individual, some common characteristics in time- and frequency-domain are observed. Figure 6 shows far-field HRIRs of KEMAR with small pinnae at horizontal azimuths 30° and 90°[8]. At azimuth 30°, the HRIR magnitude at preceding 30 to 58 samples is approximately zero, corresponding to the propagation delay from sound source to ears. In practice, a time window is usually applied to raw HRIRs, and thus the initial delay only has relative significance. The main body of the HRIRs, which reflects the complicated interactions between incident sound waves and anatomical structures, persists for about 50 to 60 samples. Subsequently, the HRIR magnitude returns to nearly zero. When the sound source deviates from directly front and back directions, the initial delay difference in the left- and right-ear HRIRs reflects the propagation time difference from the sound source to the left and right ears, i.e., ITD. At azimuth 90°, for instance, the left-ear HRIR lags to the right-ear HRIR with a relative delay of 28 samples (approximately 635 μs at a sampling frequency of 44.1 kHz). Moreover, when the sound source is located contralateral to the concerned ear, for example, at an azimuth of 90° for the left ear, the HRIR magnitude is visibly attenuated because of the head shadow effect. As elevation deviates from the horizontal plane, the difference in initial delay and magnitude between left and right HRTFs at lateral directions reduces.

Figure 7 shows the magnitudes of HRTFs corresponding to the HRIRs in Figure 6. At low frequencies below 0.4 to 0.5 kHz, the normalized log-magnitudes of HRTFs approach 0 dB and are roughly frequency-independent because of the negligible scattering and shadow effect of the head. The decrease in magnitude below 150 Hz is caused by the low-frequency limit of loudspeaker response used in HRTF measurement, rather than by the HRTF itself. Because of the finite source distance relative to the head center (r = 1.4 m) in HRTF measurement, a 2 to 4 dB difference between the left- and right-ear HRTF magnitudes is observed at a lateral azimuth of 90° even at low frequencies. As frequency increases, the normalized log-magni-

http://dx.doi.org/10.5772/56907

tudes of HRTFs vary with frequency and azimuth in a complex manner, due to the overall filtering effects of the head, pinna, torso, and ear canal. The apparent peak in HRTF magnitude at 2 to 3 kHz results from the resonance of the occluded-ear simulator of KEMAR. Above 4 kHz, the contralateral HRTF magnitudes (for example, the left ear at an azimuth of 90°) are visibly attenuated because of the low-pass filtering properties of the head shadow. The ipsilateral HRTF magnitudes (for example, the right ear at an azimuth of 90°) increase to a certain extent, although some notches occur. This phenomenon is partially attributed to the approximate mirror-reflection effect of the head on ipsilateral incidence at high frequencies, thereby leading to increased pressure for ipsilateral sound sources.

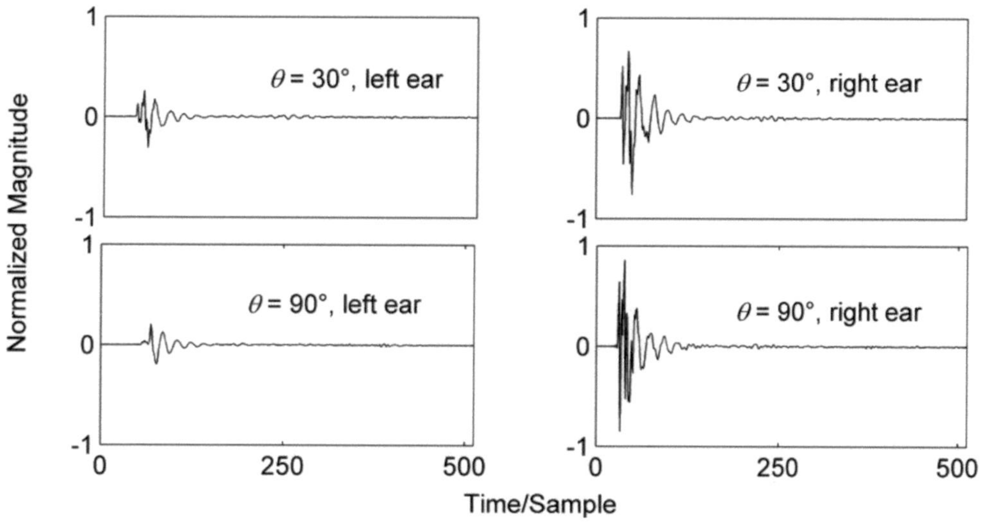

Figure 6. KEMAR far-field HRIRs at azimuths of 30° and 90° in the horizontal plane

To demonstrate the individuality of HRTFs, Figure 8 shows the normalized magnitudes of left-ear HRTFs at ($\theta = 0°$, $\phi = 0°$) for 10 subjects randomly selected from the Chinese subject HRTF database. Considerable inter-subject differences in HRTF magnitudes are observed above 6 to 7 kHz.

3.2. Localization cues in HRTFs

Various localization cues stated in Section 1 can be evaluated from measured HRTFs. ITD is a dominant azimuthal localization cue below 1.5 kHz. There are various evaluation methods for ITD, among which ITD_p calculated from interaural phase delay difference is directly related to low-frequency localization,

$$ITD_p(\theta,\phi,f) = \frac{\Delta\psi}{2\pi f} = -\frac{\psi_L - \psi_R}{2\pi f} \tag{4}$$

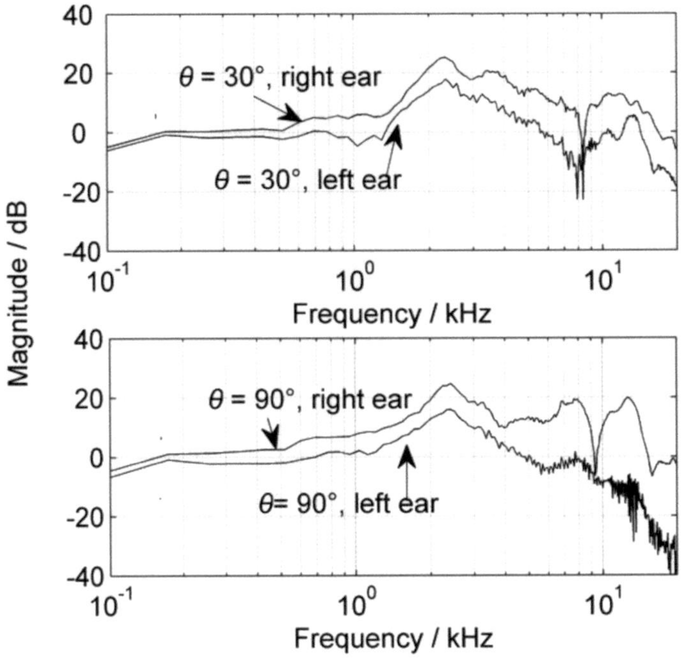

Figure 7. Magnitudes of KEMAR HRTFs at azimuths of 30° and 90° in the horizontal plane

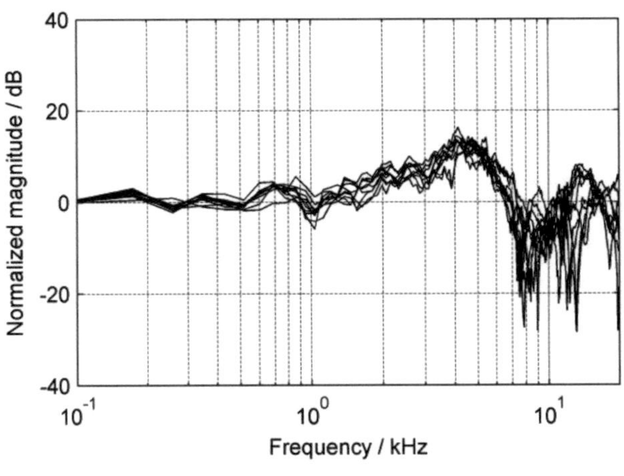

Figure 8. Left-ear HRTF magnitudes for 10 subjects at azimuth 0° in the horizontal plane

where ψ_L and ψ_R denote the unwrapped phases of left- and right-ear HRTFs, respectively. Besides, ITD can be evaluated as τ_{max} at which the normalized interaural cross-correlation function of a left- and right-ear HRIR pair maximizes.

http://dx.doi.org/10.5772/56907

$$\max\{\Phi_{LR}(\tau)\} = \max\left\{\frac{\int_{-\infty}^{+\infty} h_L(t+\tau)h_R(t)dt}{\left\{\left[\int_{-\infty}^{+\infty} h_L^2(t)dt\right]\left[\int_{-\infty}^{+\infty} h_R^2(t)dt\right]\right\}^{1/2}}\right\} \quad with\,|\tau| \leq 1ms \tag{5}$$

$$ITD_{corre}(\theta,\phi) = \tau_{\max} \tag{6}$$

In some studies, ITD is usually evaluated by leading-edge detection, i.e., detecting instants $t_{L,\eta}$ and $t_{R,\eta}$ at which the HRIRs first reach a certain percentage η (e.g., 10%) of maximum peak amplitudes. Then, ITD_{lead} is calculated by

$$ITD_{lead}(\theta,\phi) = t_{L,\eta} - t_{R,\eta}\,. \tag{7}$$

The ITD_{corre} and ITD_{lead} are relevant to source direction but independent of frequency.

Figure 9 plots the variation of horizontal ITDs with azimuths from 0° to 180°. The ITDs are calculated from MIT KEMAR (far-field) HRTFs, and left-right symmetric HRTFs are assumed. The ITDs evaluated by four different methods, including ITD_p at 0.35 and 2.0 kHz, ITD_{lead} with $\eta = 10\%$, and ITD_{corre}, are shown in the figure. Before the ITD_{corre} is calculated, a pair of HRIRs is subjected to low-pass filtering below 2.0 kHz to avoid the influence of resonance from the occluded-ear simulator. The ITDs derived by different methods generally vary with azimuth in a similar manner. The ITDs are zero at azimuths of 0° and 180°, then gradually increase as the source deviates from the median line and maximizes at directions close to the lateral. For example, the maximal ITD_{corre} is 710 μs at azimuth 90°. At a given azimuth, however, some differences in ITD value exist among the ITDs derived from different methods, with the ITD_p at 0.35 kHz being the largest and ITD_{lead} being the smallest. The range of ITD variation decreases as elevation deviates from the horizontal plane.

ILD defined in Eq. (8) is another localization cue at high frequency.

$$ILD(r,\theta,\phi,f) = 20\log_{10}\left|\frac{H_R(r,\theta,\phi,f)}{H_L(r,\theta,\phi,f)}\right| \,(dB). \tag{8}$$

According to Eq. (8), ILD depends on both source direction and frequency. Figure 10 shows ILD varying with azimuth at different frequencies. This ILD is calculated using the MIT-KEMAR (far-field) HRTFs associated with the DB-061 small pinna. At low frequency of 0.35 kHz, ILD is small (within 4.5 dB) and almost invariable with source azimuth. The non-zero ILD at low frequency is partly due to the finite source distance (1.4 m) in the MIT-KEMAR HRTF measurement. For an infinitely distant source, the ILD at low frequency trends to zero.

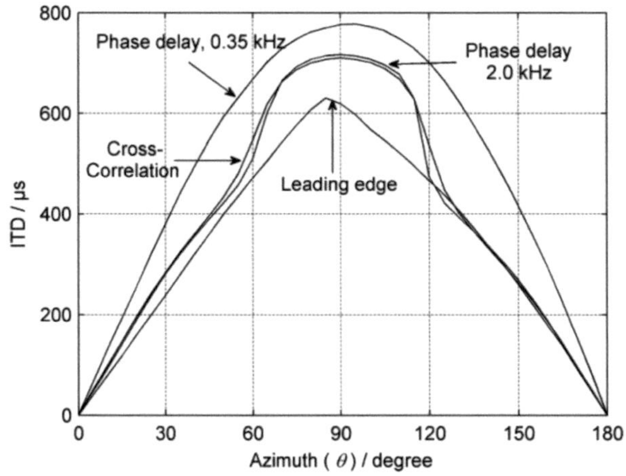

Figure 9. Horizontal ITDs of KEMAR evaluated by various methods.

As frequency increase, ILD increases and exhibits a complex variation manner with azimuth and frequency, with the value at the front (0°) and back (180°) always being zero. The range of ILD variation decreases as elevation deviates from the horizontal plane.

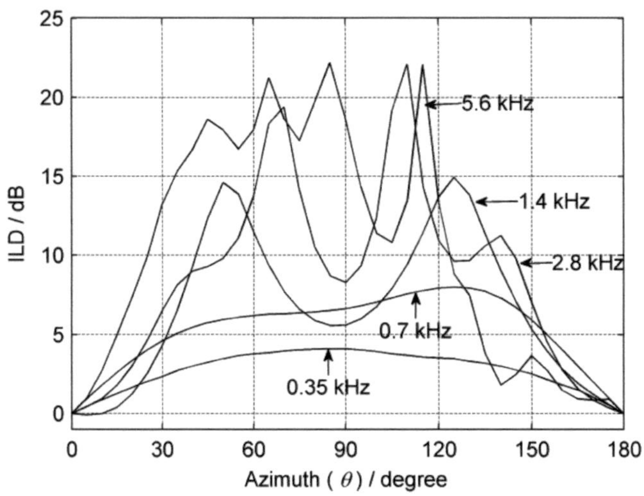

Figure 10. KEAMR ILDs in the horizontal plane for five frequencies.

The spectral cues provided by HRTFs at high frequency are vital for front-back and elevation localization. Among various spectral cues, the first (lowest) frequency notch in HRTF magnitude caused by the out-of-phase interference of pinna reflection/diffraction and direct sound wave in the ear canal is of importance. The elevation dependence of the central frequency of the pinna notch is regarded as an important vertical localization cue. Figure 11 shows the HRTF magnitude spectra of a typical Chinese subject in the median plane with $\theta = 0°$ and elevation $\phi = -30°$, 0°, and 30°[13]. The pinna notch at 6 to 9 kHz is observed in the spectra. The central frequency of the pinna notch at $\phi = -30°$, 0°, and 30° are 6.5 (6.2), 8.1(7.9), and 8.8 (8.7) kHz for the right (left) ear, respectively. At high elevations with $\phi \geq 60°$, the pinna notch gradually vanishes. Considerable inter-individual differences exist in the central frequency of the pinna notch and other high-frequency spectral features of HRTFs. Therefore, HRTFs are highly individual dependent. Actually, statistical results indicate that HRTFs are left-right asymmetric above 5 – 6 kHz [31].

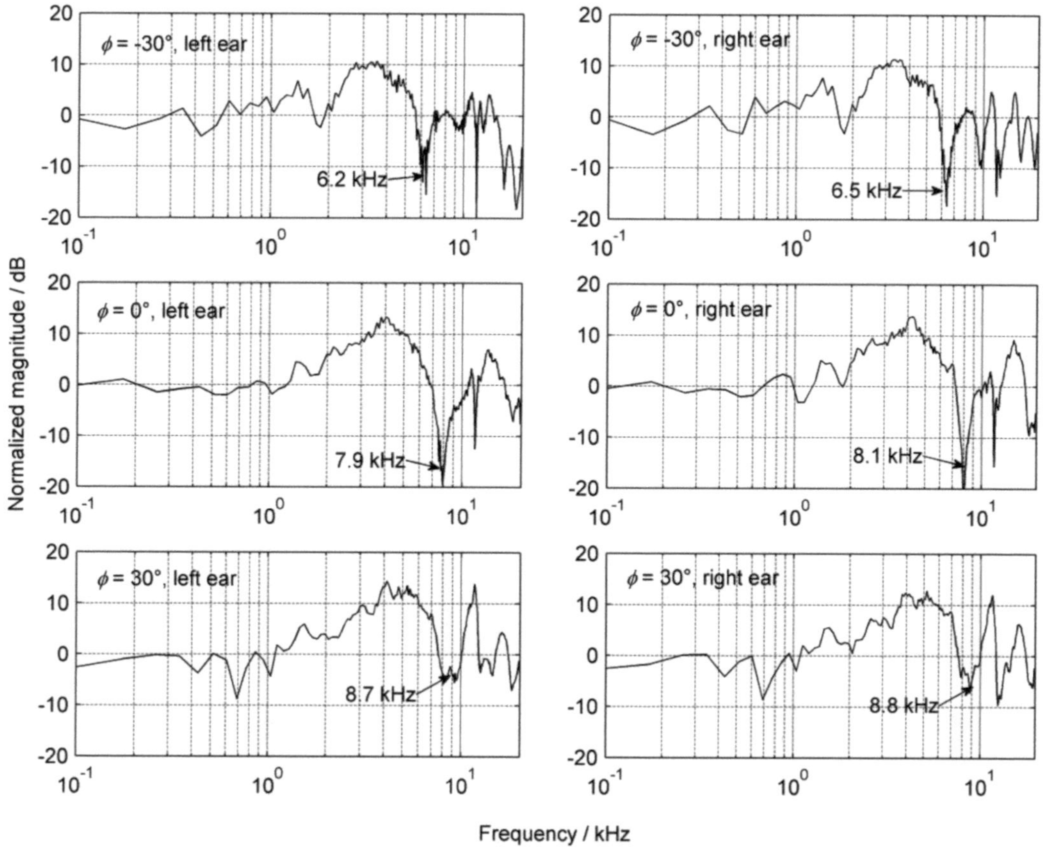

Figure 11. HRTF magnitude spectra for a typical Chinese subject at elevations –30°, 0°, and 30°

3.3. The minimum-phase characteristics of HRTFs

At a given source direction, HRTF is a complex-valued function of frequency and can be decomposed by the product of a minimum-phase function $H_{min}(\theta, \phi, f)$, an all-pass function $\exp[j\psi_{all}(\theta, \phi, f)]$, and a linear-phase function $\exp[-j2\pi f T(\theta, \phi)]$:

$$H(\theta,\phi,f) = H_{\min}(\theta,\phi,f)\exp\left[j\psi_{all}(\theta,\phi,f)\right]\exp\left[-j2\pi fT(\theta,\phi)\right] \tag{9}$$

The phase of the minimum-phase function is related to the logarithmic HRTF magnitude by Hilbert transform:

$$\psi_{\min}(\theta,\phi,f) = -\frac{1}{\pi}\int_{-\infty}^{+\infty}\frac{\ln|H(\theta,\phi,x)|}{f-x}dx\,. \tag{10}$$

If the contribution of the all-pass phase component is negligible, Eq. (9) can be approximated as

$$H(\theta,\phi,f) \approx H_{\min}(\theta,\phi,f)\exp\left[-j2\pi fT(\theta,\phi)\right]. \tag{11}$$

Eq. (11) is known as the minimum-phase approximation of HRTFs, in which an HRTF is approximated by its minimum-phase function cascaded with a linear phase or a pure delay. Studies have proved that, in most cases, HRTF is of minimum-phase below 10 – 12 kHz[32]. This conclusion is greatly convenient to the HRTF-related signal processing.

Excluding the all-pass phase component from the overall ITD calculation may cause errors when the contribution of this component is non-negligible. Minnaar et al. investigated the all-pass phase of the HRTFs of 40 subjects with 97 spatial directions per subject, and found that below 1.5 kHz the contribution of the all-pass phase component to interaural group delay difference is nearly independent of frequency[33]. If the interaural group delay difference caused by the all-pass phase component is replaced by its value at 0 Hz, the error caused by approximation is less than 30 μs and is inaudible [34].

3.4. Spatial-domain characteristics

Far-field HRTFs are continuous functions of source direction. As stated in Section 2.1, HRTFs are usually measured at discrete and finite directions, i.e., sampled at directions around a spatial spherical surface. Under certain conditions, the HRTFs at unmeasured directions (θ, ϕ) can be estimated from measured data by following linear interpolation method:

$$\hat{H}(\theta,\phi,f) \approx \sum_{i=0}^{M-1} A_i H(\theta_i,\phi_i,f)\,, \tag{12}$$

where $H(\theta_i, \phi_i, f)$ with (θ_i, ϕ_i) ($i = 0, 1, \ldots M-1$) denotes the measured HRTFs at a constant source distance $r = r_0$ and M appropriate spatial directions; A_i are a set of weights related to the target direction (θ, ϕ).

http://dx.doi.org/10.5772/56907

There are various HRTF interpolation schemes, leading to different selection of measured directions and weights. The bilinear interpolation scheme shown in Figure 12 is commonly used. Let θ_{grid} and ϕ_{grid} denote the measured intervals of azimuth and elevation, respectively. The four adjacent measured directions (θ_1, ϕ_1), $(\theta_1+\theta_{grid}, \phi_1)$, $(\theta_1+\theta_{grid}, \phi_1+\phi_{grid})$ and $(\theta_1, \phi_1+\phi_{grid})$ are denoted by number 1, 2, 3 and 4, respectively. Then the HRTF at a target direction $(\theta, \phi) = (\theta_1+\Delta\theta, \phi_1+\Delta\phi)$ within the grid is estimated as

$$\hat{H}(\theta,\phi,f) \approx A_1 H(1,f) + A_2 H(2,f) + A_3 H(3,f) + A_4 H(4,f), \tag{13}$$

where $A_\theta = \Delta\theta/\theta_{grid}$, $A_\phi = \Delta\phi/\phi_{grid}$, $A_1=(1-A_\theta)(1-A_\phi)$, $A_2=A_\theta(1-A_\phi)$, $A_3=A_\theta A_\phi$, $A_4=(1-A_\theta)A_\phi$.

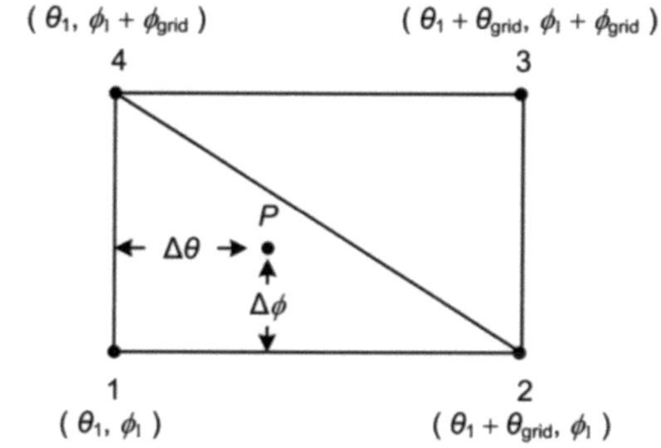

Figure 12. The bilinear interpolation

The HRTF spatial interpolation is closely related to the basis functions linear decomposition of HRTFs. HRTF linear decomposition is categorized into two basic types: spectral shape basis function decomposition and spatial basis function decomposition. Generally, the basis function decomposition representation of an HRTF for a given ear can be written as

$$H(\theta,\phi,f) = \sum_q w_q(\theta,\phi) d_q(f). \tag{14}$$

For spectral shape basis function decomposition, $d_q(f)$ are a series of frequency-dependent spectral shape basis functions; $w_q(\theta, \phi)$ are source direction-dependent weights which may also depend on individual. When the basis functions $d_q(f)$ are specified, $H(\theta, \phi, f)$ is completely determined by weights $w_q(\theta, \phi)$.

Various methods for deriving the spectral shape basis functions $d_q(f)$ are available, and appropriate selection of basis functions depends on situation. There usually exist some correlations among the HRTFs at different directions. If these correlations are completely removed so that the HRTF can be represented by a small set of spectral shape basis functions, data dimensionality is efficiently reduced. Principal components analysis (PCA) is a statistical algorithm for deriving a small set of orthonormal spectral shape basis functions and then decomposing HRTFs. For example, Kistler et al. found that five spectral shape basis functions derived from PCA accounted for more than 90% variation of logarithmic binaural HRTF magnitudes for $S = 10$ human subjects at 256 source directions[35].

In contrast, in spatial basis function decomposition, $w_q(\theta, \phi)$ in Eq. (14) denote a set of source direction-dependent spatial basis functions; $d_q(f)$ are frequency-dependent weights which may also depend on individual. There are various selections for spatial basis functions, among which azimuthal Fourier series and spatial spherical harmonic functions are two sets of pre-determined and orthonormal spatial basis functions. In the former, HRTF at each elevation plane is decomposed into a weighted sum of azimuthal harmonics. While in the latter, HRTF at arbitrary direction is decomposed into a weighted sum of spherical harmonic functions.

The spatial sampling (Shannon–Nyquist) theorem for HRTF measurement can be derived from the spatial harmonics representation of HRTF. Suppose that the spatial basis functions $w_q(\theta, \phi)$ in Eq. (14) are specified, and the basis functions up to order Q are sufficient for accurately representing HRTF. Given the measured HRTFs at M appropriate, Eq. (14) yields

$$H(\theta_i,\phi_i,f) = \sum_{q=1}^{Q} d_q(f) w_q(\theta_i,\phi_i) \qquad i = 0,1,2...(M-1)\,. \tag{15}$$

At each frequency f, Eq. (15) is a set of M linear equations, with the number of unknown $d_q(f)$ equal to the number of basis functions Q. Selecting M appropriate measurement directions and providing $M \geq Q$, the exact or approximate solution of $d_q(f)$ can be obtained from Eq. (15). The spatial basis functions representation of $H(\theta, \phi, f)$ can then be realized by substituting the resultant $d_q(f)$ into Eq. (14). Given a set of directionally continuous basis functions, HRTF at arbitrary unmeasured direction can be recovered from M directional measurements. Therefore, spatial basis functions decomposition of HRTFs can also be regarded as spatial interpolation or fitting algorithm for HRTFs. Using the azimuthal Fourier series representation of HRTF, Zhong and Xie proved that continuous HRTF in horizontal plane can be recovered from 72 azimuth measurements [36]. When extended to three-dimensional space, recovering spatial continuous HRTF using spherical harmonic functions representation requires $M = 2209$ directional measurements at least [37].

The number of directional measurements required for recovering HRTF is related to the total number of spatial basis functions (i.e., Q) for HRTF representation with $M \geq Q$. Aside from the azimuthal Fourier series and spatial spherical harmonic functions representation, if we can find another small set of spatial basis functions to efficiently represent HRTF, HRTF at unmeasured direction can be recovered from a small set of directional measurements. Xie applied spatial principal components analysis (SPCA) to a baseline HRTF dataset with high directional resolution to derive the small set of spatial basis functions[38]. SPCA is applied to spatial domain rather than frequency (or time) domain in conventional PCA. Using the resultant spatial basis functions, HRTF magnitudes at 493 directions can be recovered from 73 directional measurements. This method is applicable to simplifying HRTF measurement.

3.5. Characteristics of near-field HRTFs

When $r < 1.0$ m, the near-field HRTFs vary with source distance, and exhibit some characteristics that are remarkably distinguished from the far-field HRTFs [14, 39]. The distance dependence of near-field HRTFs is regarded as a distance perception cue. Figure 13 shows KEMAR HRTF magnitudes at $r = 0.2$ m, 0.5 m, 1.0 m and $(\theta, \phi) = (90°, 0°)$ [40]. The magnitudes vary obviously with source distance from $r = 0.2$ m to 0.5 m, and vary less with source distance from $r = 0.5$ m to 1.0 m. The ipsilateral (right) HRTF magnitude increases with decreasing r when a direct propagation path from source to concerned ear exists; the contralateral HRTF magnitude decreases with decreasing r because of the enhancement of the head shadow when a direct propagation path is missing. The variations in HRTF magnitude with r increase the ILD associated with decreasing r. This phenomenon is particularly prominent at low frequencies, thereby relatively increases low-frequency magnitude and therefore causes a perceptible change in timbre.

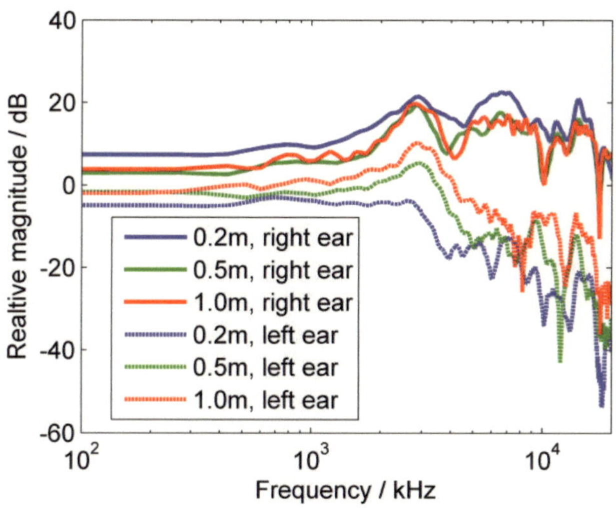

Figure 13. KEMAR HRTF magnitudes at $r = 0.2$ m, 0.5 m, 1.0 m and $(\theta, \phi) = (90°, 0°)$

4. Virtual auditory display

4.1. Basic principles

The binaural sound pressures recorded in the ear canals of a human subject or an artificial head contain the main spatial information of sound events [41]. If the eardrum pressures generated by a real sound event are replicated by sound reproduction, the same spatial auditory event or experience as the real sound event is recreated. This is the basic principle of binaural reproduction. The most straightforward method for binaural reproduction is recording the binaural signals through a pair of microphones placed in the ear canal of an artificial head or human subject and then presenting the resultant signals via a pair of headphones. This is the basic principle of the binaural recording and playback technique. Another method is synthesizing the binaural signals by signals processing and then presenting via a pair of headphone. This is the core idea of virtual auditory display (VAD).

As stated in Section 1, in the static state, the acoustical transmission from a sound source to two ears is a linear time-invariable course. In the case of free-field sound source, the binaural pressures are related to HRTFs by Eq. (1). If a mono stimulus $E_0(f)$ is filtered with a pair of HRTFs at source direction (θ, ϕ) and the resultant signals are presented via headphone, i.e.,

$$E_L(\theta,\phi,f) = H_L(\theta,\phi,f)E_0(f), \quad E_R(\theta,\phi,f) = H_R(\theta,\phi,f)E_0(f), \tag{16}$$

then the binaural pressures in reproduction is equal to or directly proportional to those created by a real source at direction (θ, ϕ), resulting in a perceived virtual source at corresponding direction. Replacing the HRTFs with different directions in Eq. (16) yields virtual sources at various directions. Note that HRTFs are individual dependent, thus an ideal VAD should use individualized HRTFs in binaural synthesis [42]. Eq.(16) can be equally expressed in the time domain as

$$e_L(\theta,\phi,t) = h_L(\theta,\phi,t) * e_0(t), \quad e_R(\theta,\phi,t) = h_R(\theta,\phi,t) * e_0(t). \tag{17}$$

That is, convoluting the mono stimulus $e_0(t)$ with a pair of HRIRs yields binaural sound signals.

4.2. Signal processing

Direct implementation of binaural synthesis in VAD by Eq. (16) or Eq. (17) usually suffers from low computational efficiency. Alternatively, various HRTF filter model and structure are often designed for binaural synthesis processing. The commonly used HRTF filter models are classified into two catalogs: the moving average (MA) model and autoregressive moving-average (ARMA) model.

In the complex-Z domain, the system function of a Q-order MA model can be written as

$$H(z) = b_0 + b_1 z^{-1} + \ldots.. + b_Q z^{-Q}, \tag{18}$$

where b_0, b_1,...b_Q are filter coefficients. In the discrete time domain, the impulse response length of a MA model is $N = Q + 1$, therefore MA is a finite impulse response (FIR) filter model.While the system function of a (Q, P)-order ARMA model can be written as

$$H(z) = \frac{b_0 + b_1 z^{-1} + ... + b_Q z^{-Q}}{1 + a_1 z^{-1} + ... + a_P z^{-P}}, \tag{19}$$

where $a_1, \ldots, a_P$ and $b_0, b_1, \ldots, b_Q$ are filter coefficients. The impulse response length of an ARMA model is infinite, therefore ARMA is an infinite impulse response (IIR) filter model.

HRTF filter design is to appropriately select the coefficients in Eq.(18) or Eq.(19) so that the filter response exactly or approximately matches the target HRTF in some mathematical or perceptual senses. Prior to filter design, some pre-processing schemes are often applied to raw HRTFs so as to simplify the resultant filters. The common simplifications include truncation by a time window so as to reduce the response length, smooth by auditory bandwidth to discard the spectral details of HRTF insignificant to auditory perception, among others. Minimum-phase approximation of HRTF is also beneficial to reduce the filter length.

Various conventional filter design methods, such as windowing or frequency sampling method for FIR filter, and Prony or the Yule–Walker method for IIR filter, have been used in HRTF filter design. Some other sophisticated methods for IIR filter design, such as balanced model truncation (BMT) [43], method using logarithmic error criterion [44] and method of common-acoustical-pole and zero [45], have also been suggested. Frequency-warped filter for HRTFs based on non-uniform frequency resolution of human hearing was also proposed [46]. Those filters can be implemented by various structures and yield reasonable physical and auditory perception performance in VAD. Reference [47] gives a review of HRTF filter design. Aside from above methods, the methods of basis functions linear decomposition of HRTFs (such as PCA) have been applied to binaural synthesis processing. The basis function decomposition-based methods allow for synthesizing multiple virtual sources with a parallel bank of common filters, and then improve the efficiency in multiple virtual source synthesis [48].

4.3. Headphone presentation

As stated in Section 2.1, the binaural signals or HRTFs can be recorded at an arbitrary reference point along the entrance of ear canal to the eardrum, or even at the blocked entrance of ear canal. Therefore, directly rendering the recorded or synthesized binaural signals via headphone without accounting for the measurement position may lead to incorrect eardrum pressures. Moreover, the non-ideal transfer characteristics of the recording and playback chain, which originates from the non-flat frequency responses of the recording microphone and reproducing headphone as well as the unwanted coupling between headphone and external ear, will inevitably cause linear distortions in both magnitude and phase of the reproduced sound pressures at the eardrums. The overall non-ideal transfer characteristics of the recording and playback chain can be represented by a pair of transfer functions, $H_{pL}(f)$ and $H_{pR}(f)$, one for each ear. Ideally, if the recorded binaural signals is equalized by the inverse of $H_{pL}(f)$ and $H_{pR}(f)$ prior to rendering to headphone, the linear frequency distortion in the signal chain can then be eliminated or at least reduced as minimally as possible.

$$F_L(f) = \frac{1}{H_{pL}(f)} \quad and \quad F_R(f) = \frac{1}{H_{pR}(f)}. \tag{20}$$

Figure 14 is the blocked diagram of binaural synthesis along with headphone equalization in a VAD.

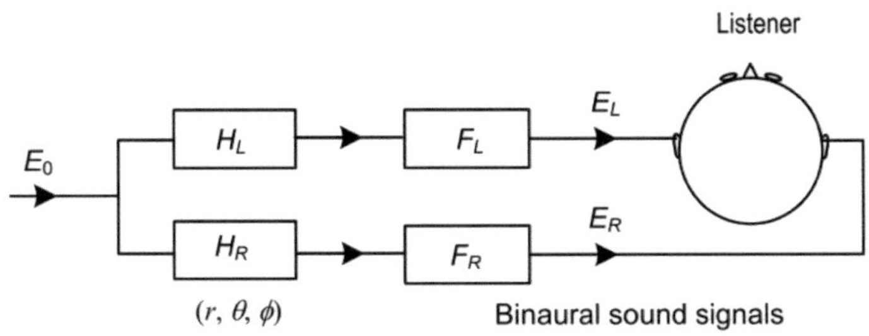

Figure 14. The blocked diagram of binaural synthesis along with headphone equalization

In particular, the transmission character from the electric input signal of headphone to the reference point in the ear canal is described by headphone-to-ear-canal transfer function (HpTF). If the reference point of HpTFs are identical to that of HRTFs and microphones for binaural recording or HRTF measurement have an ideal transmission response, the $H_{pL}(f)$ and $H_{pR}(f)$ in Eq. (20) can be replaced by HpTF, i.e., the binaural signals are equalized by the inverse of HpTFs. For microphone with non-ideal transmission response, providing that the microphones used in HpTFs measurement are identical to those in binaural recording or HRTF measurement, the effect of the non-ideal transmission response of microphone is cancelled in equalization [2]. Note that, for a blocked ear canal measurement, the above equalization method is not always valid unless a headphone with free-air equivalent coupling to the ear (FEC-hedphone) is used. The transmission response on 14 types of headphones were measured [49], and results indicated that the responses of all the headphones (except one) deviated from that of ideal FEC-headphone on the order of 2 to 4 dB above 2 kHz. Moreover, the measurements above 7 kHz were unreliable. In practical uses, whether a headphone can be considered as an FEC-headphone depends on acceptable error.

Similar to the case of HRTFs, HpTFs is individual dependent because of the difference in structures and dimensions of the external ear. Ideally, individualized HpTFs should be incorporated into equalization processing. Moreover, the measured HpTFs for some types of headphone exhibit poor repeatability above 5 to 6 kHz due to the variation of compressive deformation of pinna caused by headphone. This phenomenon makes the equalization difficult.

In headphone presentation, an accurate virtual source can be rendered if the sound pressures for a real sound source are exactly replicated at eardrums. Results of some psychoacoustic experiments with careful individualized HRTFs processing and HpTFs equalization indicate that headphone-rendered virtual source could achieve the equivalent localization performance as that of free-field real source [50]. However, numerous experimental results indicate that subject-dependent errors in perceived virtual source position are generally existed such as

1. Reversal Error (i.e., front-back or back-front confusion). That is, a virtual source intended in the front hemisphere is perceived at a mirror position in the rear hemisphere, or, less frequently, the reverse. Sometimes, there is confusion with up and down source positions termed up-down or down-up confusion.

2. Elevation error. For example, the direction of a virtual source in the front median plane is usually elevated.

3. In-head localization (i.e., intracranial lateralization). The virtual source or auditory event is perceived inside the head rather than outside headphone, leading to an unnatural hearing experience.

As stated in Section 1, the interaural cues such as ITD and ILD only determine a confusion cone rather than a well-defined spatial position of sound source. The dynamic cue caused by head movement and high-frequency spectral cue introduced by pinnae etc. response for resolving reversal ambiguation and vertical localization. However, conventional static VAD is lack of dynamic cues, so that front-back and vertical localization depend more on high-frequency spectral cue. Unfortunately, the high-frequency spectral cue is elaborate and highly individual-dependent. Errors in binaural recording/synthesis and playback chain, such as non-individualized HRTFs processing, incorrect or lack of headphone equalization, are possible sources responsible for perceived position errors in headphone presentation. Using individual HRTFs and HpTFs processing reduces localization errors. In addition, modeling room reflections in binaural synthesis effectively eliminates in-head localization.

4.4. Loudspeaker presentation

Binaural signals from either binaural recording or synthesis, are originally intended for headphone presentation. When binaural signals are reproduced through a pair of left and right loudspeakers arranged in front of the listener, an unwanted cross-talk from each loudspeaker to the opposite ear occurs. Cross-talk impairs the directional information encoded in the binaural signals. Therefore, cross-talk cancellation should be introduced for binaural reproduction through loudspeakers [51]. That is, prior to loudspeaker reproduction, binaural signals should be pre-corrected or filtered so as to cancel the transmission from each loudspeaker to the opposite ear.

Let $E_L(f)$ and $E_R(f)$, or simply E_L and E_R, denote frequency-domain binaural signals. As illustrated in Figure 15, binaural signals are pre-filtered by a 2×2 cross-talk cancellation matrix and then reproduced through the loudspeakers. The loudspeaker signals are given by

$$\begin{bmatrix} L' \\ R' \end{bmatrix} = \begin{bmatrix} A_{11} & A_{12} \\ A_{21} & A_{22} \end{bmatrix} \begin{bmatrix} E_L \\ E_R \end{bmatrix} \tag{21}$$

where A_{11}, A_{12}, A_{21} and A_{22} are the four transfer functions or filters forming the cross-talk cancellation matrix.

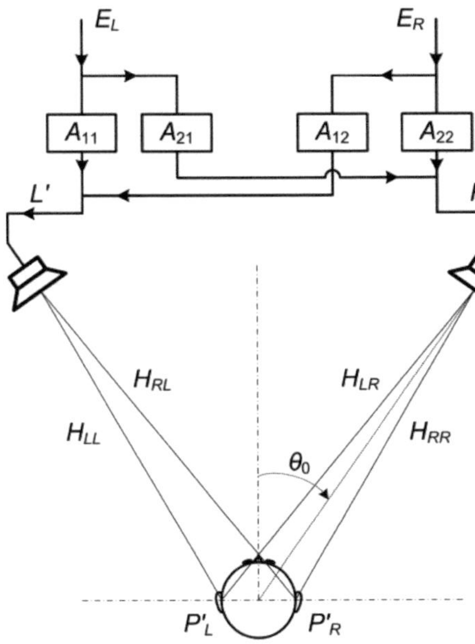

Figure 15. Binaural reproduction through loudspeakers

Let H_{LL}, H_{RL}, H_{LR} and H_{RR} denote the four acoustic transfer functions (HRTFs) from two loudspeakers to two ears, respectively. These four transfer functions are determined by the loudspeaker configuration and listener's location. Then the reproduced pressures at two ears are given by

$$\begin{bmatrix} P_L' \\ P_R' \end{bmatrix} = \begin{bmatrix} H_{LL} & H_{LR} \\ H_{RL} & H_{RR} \end{bmatrix} \begin{bmatrix} L' \\ R' \end{bmatrix} = \begin{bmatrix} H_{LL} & H_{LR} \\ H_{RL} & H_{RR} \end{bmatrix} \begin{bmatrix} A_{11} & A_{12} \\ A_{21} & A_{22} \end{bmatrix} \begin{bmatrix} E_L \\ E_R \end{bmatrix} \tag{22}$$

with Eq. (21) substituted. The transfer characters of cross-talk cancellation matrix are properly selected so that the product of two 2×2 matrixes in Eq. (22) equals to an identity matrix, then the cross-talk is completely cancelled out and the desired binaural signals are exactly delivered to listener's two ears. Therefore, the cross-talk cancellation matrix is obtained from the inverse of the acoustic transfer matrix. In the left-right symmetrical case, $H_{LL} = H_{RR} = H_\alpha$ and $H_{LR} = H_{RL} = H_\beta$, the element of cross-talk cancellation matrix is

$$A_{11} = A_{22} = \frac{H_\alpha}{H_\alpha^2 - H_\beta^2}, \qquad A_{12} = A_{21} = \frac{-H_\beta}{H_\alpha^2 - H_\beta^2}. \tag{23}$$

http://dx.doi.org/10.5772/56907

If the signal processing initially aims to create appropriate loudspeakers signals, the two stages of binaural synthesis and cross-talk cancellation can be merged together, yielding

$$L' = G_L(\theta, f)E_0(f), \qquad R' = G_R(\theta, f)E_0(f). \tag{24}$$

with

$$G_L(\theta, f) = \frac{H_\alpha H_L(\theta, f) - H_\beta H_R(\theta, f)}{H_\alpha^2 - H_\beta^2}, G_R(\theta, f) = \frac{-H_\beta H_L(\theta, f) + H_\alpha H_R(\theta, f)}{H_\alpha^2 - H_\beta^2}. \tag{25}$$

Eq. (24) demonstrates that loudspeaker signals L' and R' for target virtual source at direction θ can be directly synthesized by filtering a mono stimulus $E_0(f)$ with a pair of filters $G_L(\theta, f)$ and $G_R(\theta, f)$. This is the basic principle of loudspeaker-based binaural reproduction or transaural synthesis. The cross-talk cancellation and transaural synthesis can be generalized to the case of binaural reproduction through more than two loudspeakers and with more than one listener [52]. In practice, the acoustic transfer matrix in Eq. (24) may be singular and thus non-invertible at some frequencies. To address this problem, some approximate methods for solving cross-talk cancellation matrix have been proposed [53].

The first problem with binaural reproduction through two frontal loudspeakers is reversal and elevation errors. High-frequency spectral cues is vital to front-back and vertical localization. But these cues cannot be stably replicated in loudspeaker reproduction because of the short wavelength at high frequency. A slight change in listening position causes an obvious variation in binaural pressures. Incorrect dynamic cues often causes back-front confusion in static binaural reproduction through a pair of frontal loudspeakers. In contrast to headphone reproduction, two-front loudspeaker reproduction can recreate only stable perceived virtual sources in frontal-horizontal quadrants rather than in full three-dimensional directions.

The second problem with loudspeaker reproduction is a limited listening region or sweet points. For a given loudspeaker configuration, the cross-talk in Eq. (23) or transaural synthesis in Eq. (25) is designed according to a default (optimal) listening position. Head deviation from the default position (including translation and tuning) spoils the cross-talk cancellation, and thus alters the binaural pressures. Therefore, the performance of cross-talk cancellation is position-dependent. There have been a lot of works on the stability of loudspeaker-based binaural reproduction against head movement [54-57]. Kirkeby *et al.* proved that two frontal loudspeakers configuration with narrow span angle is beneficial to the stability of virtual source [54, 55]. Kirkeby further proposed using a pair of frontal loudspeakers with 10° span (in contrast to 60° span in conventional stereo) for binaural or transaural reproduction, which is known as "stereo dipole". A stereo dipole improves the stability of virtual source at mid-frequency at the cost of making low-frequency signal processing difficult because a large low-frequency boost is required.

The third problem with loudspeaker reproduction is timbre coloration. Ideally, a perfect cross-talk cancellation yields the same binaural pressures as those with a real source. Nevertheless, as stated above, it is difficult to cancel out cross-talk completely within a full audible frequency range. In practice, some reasons, such as slight movement, unmatched HRTFs, and room reflection etc., inevitably lead to incomplete cross-talk cancellation so that the binaural pressures at reproduction deviate from those of a real source. This in turn leads to perceived coloration, especially at high frequency and for an off-center listener. Therefore additional timbre equalization is required.

The principle of timbre equalization in two frontal loudspeakers reproduction can be explained as follows. Due to the difficulty in robust rendering the fine high-frequency spectral cues to listener's ears in loudspeaker reproduction, the perceived virtual source direction is dominated by the interaural cues (especially ITD) and limited to the frontal horizontal quadrant. While the interaural cues are controlled by the relative rather than the absolute magnitude and phase between left and right loudspeaker signals. Scaling both loudspeaker signals with identical frequency-dependent coefficient does not alter their relative magnitude and phase and thus the perceived virtual source direction. However, this manipulation alters the overall power spectra of the loudspeaker signals and thus impairs the timbre. Xie *et al.* proposed a constant-power equalization algorithm, in which the responses of transaural synthesis filters $G_L(\theta, f)$ and $G_R(\theta, f)$ in Eq. (24) were equalized by their root-mean-square [58]. As a result, the $G_L(\theta, f)$ and $G_R(\theta, f)$ in Eq. (25) are replaced by

$$\begin{aligned} G_L'(\theta,f) &= \frac{G_L(\theta,f)}{\sqrt{|G_L(\theta,f)|^2+|G_R(\theta,f)|^2}} \\ &= \frac{H_\alpha H_L - H_\beta H_R}{\sqrt{|H_\alpha H_L - H_\beta H_R|^2 + |-H_\beta H_L + H_\alpha H_R|^2}} \frac{|H_\alpha{}^2 - H_\beta{}^2|}{H_\alpha{}^2 - H_\beta{}^2}; \\ G_R'(\theta,f) &= \frac{G_R(\theta,f)}{\sqrt{|G_L(\theta,f)|^2+|G_R(\theta,f)|^2}} \\ &= \frac{-H_\beta H_L + H_\alpha H_R}{\sqrt{|H_\alpha H_L - H_\beta H_R|^2 + |-H_\beta H_L + H_\alpha H_R|^2}} \frac{|H_\alpha{}^2 - H_\beta{}^2|}{H_\alpha{}^2 - H_\beta{}^2}. \end{aligned} \tag{26}$$

With $G_L'(\theta, f)$ and $G_R'(\theta, f)$, it can be proved that the loudspeaker signals given by Eq. (24) satisfy following relationship of constant power spectra:

$$|L'|^2 + |R'|^2 = E_0^2 \tag{27}$$

Therefore, the overall power spectra of loudspeaker signals is equal to that of the input stimulus, so reproduction coloration reduces.

http://dx.doi.org/10.5772/56907

4.5. Simulation of reflections

Free-field virtual source synthesis and rendering are discussed above, in which room or environment reflections are ignored. However, reflections exist in most real rooms and are vital to spatial auditory perception. Therefore, a complete VAD should include reflection modeling, and thereby is called virtual auditory or acoustic environment (VAE). Incorporating reflections into VAE processing brings following advantages: (1) recreating the spatial auditory perception in a room or reflective environment; (2) eliminating or reducing the in-head localization in headphone presentation; (3) controlling perceived virtual source distance.

Usually, there are two basic methods for room or environment reflection rendering. The physics-based method simulates the physical propagation of sound from source to receiver inside a room, or equally, the binaural room impulse responses (BRIRs), and then synthesizes the binaural signals by convoluting the input stimulus with BRIRs. The perception-based method recreates desired auditory perception of reflections by some signal processing algorithms from perceptual rather than physical viewpoint.

A complete physical modeling of BRIRs consists of source modeling (such as source radiation pattern), transmission or room acoustics modeling (such as frequency-dependent surface reflection, scattering and absorption, air absorption, etc.), listener modeling (scattering and diffraction by human anatomical structures). The room acoustics modeling methods are divided into two categories according to physical principle, i.e., geometrical acoustics-based method and wave acoustics-based method. Geometrical acoustics neglects most wave nature of sound, yielding the approximate solutions of room acoustic field. This approximation is reasonable for high frequency and smooth boundary surface. The image-source method and ray-tracing method are two commonly used geometrical acoustics-based methods. The former decomposes the reflection sound field into the radiations of multiple image sources in free space. While the later treats sound radiation like a number of rays, which propagate and then are reflected and absorbed by boundary surface according to certain rule.

When the wave nature of sound is taken into account, wave acoustics-based methods should be used. These methods solve the wave equation for pressure inside the room and yield more accurate results. Various numerical methods, such as the finite element method, boundary element method, finite-difference time domain method and digital waveguide mesh method, have been suggested to solve the acoustic field in rooms with complex geometries. Limited to the extensive computational workload, however, these numerical methods are merely suitable for low-frequency and small room modeling.

Room acoustic field modeling yields time, direction, magnitude (or energy) as well as the spectra of each reflection arriving at a received point. Each reflection is filtered with a pair of corresponding HRTFs and the contribution of all reflections are combined to form complete BRIRs. In actual VAD or VAE, convolution of the input stimulus with HRIRs can be implemented by some decomposed structures.

The calculation for modeling and convoluting with a pair of complete BRIRs is complex. In some practical uses, the physics-based methods mentioned above are used to simulate and render the early room reflections in VAEs. To simplify processing, the late and diffuse room reflections are often simulated by some perception-based methods, such as various artificial delay and reverberation algorithms [59]. These algorithms are based on the pre-measured or pre-calculated room acoustic attributes or parameters (such as reverberation time) and render the reflections from the perceptual rather than physical point of view.

4.6. Dynamic VAD

In static VAD or VAE discussed above, both virtual sources and listeners are assumed to be fixed and real-time processing is not always required. In a real acoustic environment, however, either source or listener movement alters the binaural pressures and brings dynamic acoustic information. This dynamic information should be incorporated into VAD or VAE processing, because it is significant for both source localization and recreating convincing auditory perceptions of acoustic environment. Therefore, in addition to modeling the sound source, room (environment) and listener, a sophisticated VAD should be able to constantly detect the position and orientation of listener's head, based on which the signal processing is updated in real-time. In other words, a faithful VAE should be an interactive, dynamic and real-time rendering system, and thus called dynamic and real-time VAD system.

Figure 16 shows the basic structure of a dynamic VAD system, which consists of three parts:

1. Information input and definition

This part inputs the prior information and data for dynamic VAD through a user interface. These information and data are classified into three categories: source information, environment information and listener information. The source information includes type of source stimuli, the number, spatial positions, orientation, directivities (radiation pattern) and level of sources, or predetermined trajectory for a moving source, etc. The environment information includes room or environment geometry, absorption coefficients of surface material and air, etc. The listener information includes the initial spatial position, orientation and individual data of listener (such as HRTFs). A head-tracking device detects the position and orientation of listener's head and then provides those information to the system.

2. Dynamic VAD signal processing

According to prior information and data in part 1, this part simulates sound source as well as both direct and reflected/scattered propagation from sound sources to two ears using certain physical algorithms. Based on the temporary position of the head detected by head-tracking devices, the HRTFs for binaural synthesis are constantly updated so as to obtain dynamic binaural signals.

3. Reproduction

The resultant binaural signals are reproduced through headphone after headphone equalization, or through loudspeakers after cross-talk cancellation.

Ideally, the binaural signals or auditory scenario created by a dynamic VAD should synchronously vary with head movement just as in the real environment. Therefore, an ideal dynamic VAD should be a linear time-variable system. However, the signal processing schemes in

http://dx.doi.org/10.5772/56907

dynamic VAD are deduced from the static scheme, in which a series of short "static state" are used for approximating the transient. Therefore, the dynamic behaviors of VAD should be considered.

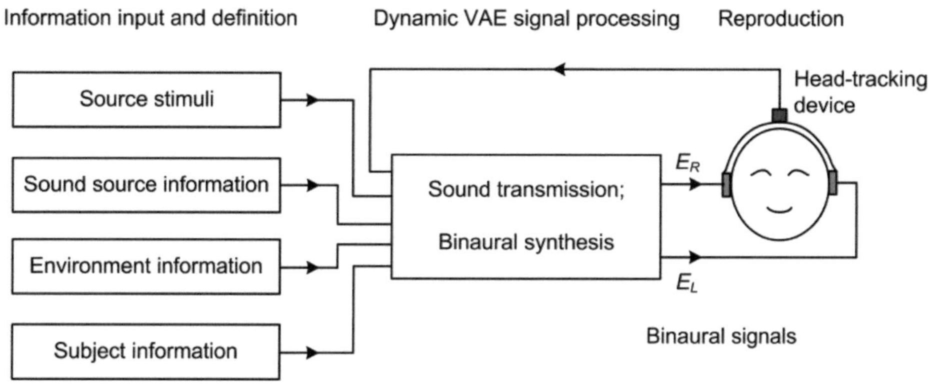

Figure 16. Structure of a typical dynamic VAD system

One problem concerned with the dynamic behaviors is the scenario update rate. A dynamic VAD updates the binaural signals and thereby auditory scenario at certain time interval. The scenario update rate of a VAD refers to the number of update scenario manipulations per second. Another problem concerned with dynamic behaviors is the system latency time. When the head moves, the synthesized binaural signals in existing VAD do not change synchronously but with a delay. The system latency time refers to the time from the listener's head movement to corresponding change in the synthesized binaural signals output, which is contributed by the hardware (such as head tracker) and software structures, as well as the data transmission and communication of the system. Usually, a high scenario update rate and low system latency time are preferred for improving the performance of a dynamic VAD system. Limited by the available system capacity, however, some tradeoffs in system performance should be made in practical dynamic VAD on basis of psychoacoustic rules. In addition, the auditory continuity of scenario update should also be taken in account.

Some psychoacoustic experiments have been carried out to investigate the required scenario update rate and system latency time in a dynamic VAD. The experiment conducted by Sandvad indicated that a scenario update rate of 10 Hz or less degraded the speed of localization; and a scenario update rate of 20 Hz almost would not degrade the speed of localization, although audible artifacts may occur for moderate to fast head movement [60]. While the results for system latency time vary among different studies [61-63]. A general accepted conclusion is that a system latency time lower than 60 ms is adequate for most applications [63].

Some dynamic VAD systems have been developed for various purposes and applications [64-71]. For example, the SLAB (Sound Laboratory) developed by NASA in U. S. was intended to provide a platform for psychoacoustic study [64-66]. It is a software-based system written with VC++ language, and is implemented by a PC or server under Microsoft Windows operating system. Through application programming interface (API), the SLAB provides access to different psychoacoustic researches. In the SLAB, the dynamic virtual auditory environment caused by moving sources in real-time can be rendered, including the simulation of source radiation pattern, sound propagation, environment reflection and absorption, air absorption etc. Six of 1st-order reflections were modeled by image source method. The maximum achievable number of virtual sources depends on the available computational ability of system (typical 4 CPUs). The typical scenario update rate is 120 Hz. Excluding the external latency caused by head tracker, the internal system latency time is 24 ms. The binaural signals are reproduced through headphone. The SLAB system has been updated for several times. The latest version also supports using individualized HRTFs.

5. Applications of VADs

5.1. Psychoacoustic experiment and hearing training

Psychoacoustic acoustics investigates the relationship between acoustics-related physical factors and resulting subjective perceptions. By means of VADs, the complete and precise controlling over some physical characteristics of binaural signals is allowed, and corresponding subjective perceptions can be created. Therefore, VADs have become an important experimental tool for psychoacoustic acoustics, such as auditory localization mechanism [72] and masking [73]. VADs also benefit to hearing training for musicians and sound engineers.

5.2. Virtual reality and multimedia

Virtual reality is a kind of human-computer interface technology that provides users the feeling of being presence by including various perceptual cues such as visual, auditory, tactile sense [68]. The interaction and complementary of multiple information on above aspects strengthen the sense of reality and immerse. By means of VADs, various auditory perceptions to source localization or acoustical environments can be generated. Therefore, VADs are important to virtual reality in regard to auditory simulation. A typical example is driving training simulation [74], which can be realized by the dynamic VADs presented in Section 4.6. Similar methods can also be applied to some special environment trainings such as virtual aviation, aerospace, submarine environments.

VADs have been widely applied on the entertainment functions of multimedia PC. At present, game softwares under Windows platform possess the functions of VADs. In such kind of consumer electronics, simplified signal processing in VADs is needed in consideration of cost and computer capacity.

http://dx.doi.org/10.5772/56907

5.3. Speech communication

Psychoacoustic research indicates that the target detection ability of the binaural hearing is prior to the monoaural hearing in presence of background interference. In daily life, a listener can detect the target speech information even in a noisy environment with negative signal-to-noise ratio, suggesting the high speech intelligibility of the target. This is so-called the cocktail party effect [75]. However, mono signal transmission is dominant in currently available communication systems, resulting in low speech intelligibility. This condition can be improved by using binaural signal transmission, in which spatial separation between target and competing sources is realized through VADs. This method can be applied to teleconference and other speech communication systems. VADs are also helpful in aeronautical communication on aspects of improving speech intelligibility and reducing the react time of the pilot in the case of accident hazard [76].

5.4. Binaural auralization

On-site listening is the most straightforward way for subjective assessment of room acoustic quality. However, this is difficult in practical use. One reason is the impossibility of accurately compare among halls at different areas due to human short-term memory and expansive travelling cost. Moreover, it is difficult to orginize the same band to play the same music at different halls, which is needed in accurate subjective assessment.

As mentioned, BRIRs contain the main information of direct sounds and reflections. Binaural auralization is achieved by convoluting the mono "dry" signal with mathematically or physically-obtained BRIRs (see Section 4.5) and reproducing the synthesized binaural signals through headphone or louderspeakers with proper crosstalk cancelling. In past decades, binaural auralization has become an important tool in the research and design of room acousitc quality [77]. Especially, binaural auralization is helpful to detect acoustic defects in regard to subjective properties on the stage of room design. This function has been included in some softwares for room acoustic design such as Odeon. Besides, binaural auralization has been generally used in subjective assessment such as noise evaluation [78], subjective assessment of sound reproduction systems [79], and virtual sound recording.

5.5. Virtual reproduction for multi-channel surround sound

Multi-channel surround sound reproduction, such as 5.1 channel surround sound, requires multiple loudspeakers, which is complex and inconvenient in some practical applications such as TV or multimedia computer. Hence, some HRTF-based virtual loudspeaker-based approaches (i.e., HRTF-based binaural synthesis in Section 4.4) for multi-channel surround sound reproduction have been introduced to reduce the number of loudspeaker needed. For example, some commercial products have been introduced for the virtual reproduction of 5.1 channel surround sound, that is, simulating 5.1 channel surround sound through a pair of actual stereophonic loudspeakers, see Figure 17. Signals L and R are directly fed to left and right loudspeakers respectively so as to create summing virtual source within the span of two loudspeakers. Signal C is attenuated 3 dB and then fed to the left and right loudspeakers to create a summing virtual source at the front $\theta = 0°$. Two surround signals *LS* and *RS* are filtered by transaural synthesis filters and then fed to the loudspeakers.

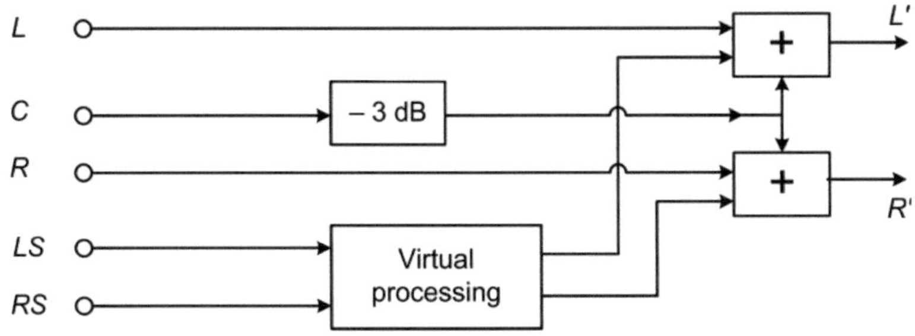

Figure 17. The block diagram of virtual 5.1 channel surround sound through loudspeakers

HRTF-based binaural synthesis can also be applied into multi-channel surround sound virtual reproduction through headphone. When directly rendering multi-channel surround sound signals to headphone, unnatural perceptions such as in-head localization occur. Using HRTF-based binaural synthesis, multiple loudspeakers can be virtually generated by headphone. Correspondingly, subjective perception of headphone-based multi-channel surround sound reproduction can be improved. Some related products have been introduced, such as Dolby headphone (http://www.dolby.com).

In virtual reproduction for multi-channel surround sound, defects presented in Sections 4.3 and 4.4 for VADs also exist, such as timbre coloration, limited listening area, and directional distortions. Our group has proposed some patents on the improvement of those defects.

6. Summary

HRTFs capture most localization information in binaural pressures and exhibit important physical and auditory characteristics. One major application of HRTFs is VAD or VAE, in which HRTF-based signal processing is used to recreate virtual source and other spatial auditory events in headphone presentation or loudspeaker presentation with proper equalization. Great developments have been achieved in the field of HRTFs and VADs, but many issues need further research. VADs have currently been applied to various fields in scientific research, engineering, entertainment and consumer electronic products, among others.

Acknowledgements

This work is supported by the National Nature Science Foundation of China (Nos. 11174087, 11004064), and State Key Lab of Subtropical Building Science, South China University of Technology.

http://dx.doi.org/10.5772/56907

Author details

Xiao-li Zhong and Bo-sun Xie

Acoustic Lab, Physics Dept., School of Science, South China University of Technology, Guangzhou, China

References

[1] Blauert J. Spatial Hearing (Revised edition), MIT Press, Cambridge, MA, England, 1997

[2] Møller H. Fundamentals of binaural technology. Applied Acoustics, 1992; 36(3/4), 171-218

[3] Xie B S. Head Related Transfer Function and Virtual Auditory Display. USA: J.Ross Publishing, 2013

[4] Xiang N, Schroeder M R. Reciprocal maximum-length sequence pairs for acoustical dual source measurements. J. Acoust. Soc. Am., 2003; 113 (5), 2754-2761

[5] Yu G Z, Liu Y, Xie B S. Fast measurement system and super high directional resolution head-related transfer function database. J. Acoust. Soc. Am.,2012; 131(4), 3304

[6] Wightman F L, Kistler D J. Measurement and validation of human HRTFs for use in hearing research. Acta Acoustica United with Acoustica, 2005; 91 (3), 429–439

[7] Blauert J, Brueggen M, Bronkhorst A W, et al. The AUDIS catalog of human HRTFs. J. Acoust. Soc. Am., 1998; 103 (5), 3082

[8] Gardner W G, Martin K D. HRTF measurements of a KEMAR. J. Acoust. Soc. Am., 1995; 97 (6), 3907-3908

[9] Møller H, Sørensen M F, Hammershøi D, et al. Head-related transfer functions of human subjects. J.Audio Eng. Soc., 1995; 43(5), 300-321

[10] Algazi V R, Duda R O, Thompson D M, et al. The CIPIC HRTF database, Proceeding of 2001 IEEE Workshop on the Applications of Signal Processing to Audio and Acoustics, 99 -102, 2001

[11] IRCAM Lab (2003): Listen HRTF database, http://recherche.ircam.fr/equipes/salles/listen/ (accessed 25 Aug. 2013)

[12] Genuit K, Xiang N. Measurements of artificial head transfer functions for auralization and virtual auditory environment. Proceeding of 15th International Congress on Acoustics (invited paper), Trondheim, Norway, II 469-472 ; 1995

[13] Xie B S, Zhong X L, Rao D, et al. Head-related transfer function database and analyses. Science in China Series G, Physics, Mechanics & Astronomy. 2007; 50(3), 267-280

[14] Brungart D S, Rabinowitz W M. Auditory localization of nearby sources. Head-related transfer functions. J. Acoust. Soc. Am., 1999; 106 (3), 1465-1479

[15] Hosoe S, Nishino T, Itou K, et al. Measurement of Head-related transfer function in the proximal region. Proceeding of Forum Acusticum 2005, Budapest, Hungary, 2539-2542,2005

[16] Gong M, Xiao Z, Qu T S, et al. Measurement and analysis of near-field head-related transfer function, Applied Acoustics (in Chinese). 2007; 26, 326—334

[17] Yu G Z, Xie B S, Rao D. Characteristics of Near-field head-related transfer function for KEMAR. AES 40th Conference. Japan, Tokyo; 2010

[18] Morse P M, Ingrad K U. Theoretical Acoustics, McGraw-Hill, New York, USA, 1968

[19] Duda R O, Martens W L. Range dependence of the response of a spherical head model. J.Acous.Soc.Am.,1998; 104(5), 3048-3058

[20] Algazi V R, Duda R O, Duraiswami R, et al. Approximating the head-related transfer function using simple geometric models of the head and torso. J.Acoust.Soc.Am., 2002; 112 (5), 2053-2064

[21] Gumerov N A, Duraiswami R. Computation of scattering from N spheres using multipole reexpansion. J. Acoust. Soc. Am., 2002; 112 (6), 2688-2701

[22] Kahana Y, Nelson P A. Boundary element simulations of the transfer function of human heads and baffled pinnae using accurate geometric models. J. Sound and Vibration, 2007; 300(3/5), 552-579

[23] Katz B F G. Boundary element method calculation of individual head-related transfer function.I. Rigid model calculation. J.Acoust.Soc.Am., 2001; 110(5), 2440-2448

[24] Otani M, Ise S. Fast calculation system specialized for head-related transfer function based on boundary element method. J.Acoust.Soc.Am.,2006; 119(5), 2589-2598

[25] Gumerov N A, O'Donovan A E, Duraiswami R, et al. Computation of the head-related transfer function via the fast multipole accelerated boundary element method and its spherical harmonic representation. J. Acoust. Soc. Am., 2010; 127(1), 370-386

[26] Zotkin D N, Hwang J, Duraiswami R, et al. HRTF personalization using anthropometric measurements, Proceedings of the 2003 IEEE Workshop on Applications of Signal Processing to Audio and Acoustics, 157-160; 2003

[27] Middlebrooks J C. Individual differences in external-ear transfer functions reduced by scaling in frequency, J. Acoust. Soc. Am., 1999; 106 (3), 1480-1492

[28] Nishino T, Inoue N, Takeda K, et al. Estimation of HRTFs on the horizontal plane using physical features, Applied Acoustics, 2007; 68(8), 897-908

[29] Seeber B, Fastl H. Subjective selection of non-individual head-related transfer functions, Proc. ICAD 2003, 259–262; 2003

[30] Yairi S, Iwaya Y, Suzuki Y. Individualization feature of head-related transfer functions based on subjective evaluation, Proceedings of the 14 International Conference on Auditory Display, Paris, France June 24 – 27; 2008

[31] Zhong X L, Xie B S. Spatial symmetry of head-related transfer function, Chinese Journal of Acoustics, 2007; 26, 73—84

[32] Kulkarni A, Isabelle S K, Colburn H. S. Sensitivity of human subjects to head-related transfer-function phase spectra. J. Acoust. Soc. Am., 1999; 105,2821—2840

[33] Minnaar P, Christensen F, and Møller H, et al. Audibility of all-pass components in binaural synthesis. AES 106th Convention, Munich, Germany, Preprint: 4911, 1999.

[34] Plogsties J, Minnaar P, and Olesen S, et al. Audibility of all-pass components in head-related transfer functions. AES 108th Convention, Paris, France, Preprint: 5132, 2000

[35] Kistler D J, Wightman F L. A model of head-related transfer functions based on principal components analysis and minimum-phase reconstruction. J. Acoust. Soc. Am., 1992; 91,1637-1647

[36] Zhong X L, Xie B S. Maximal azimuthal resolution needed in measurements of head-related transfer functions. J. Acoust. Soc. Am.,2009; 125,2209—2220

[37] Zhang W, Zhang M Q, Kennedy R A, et al. On high-resolution head-related transfer function measurements: an efficient sampling scheme. IEEE Transactions on Audio, Speech, and Language Processing, 2012; 20(2):575—584

[38] Xie B S. Recovery of individual head-related transfer functions from a small set of measurements, J. Acoust. Soc. Am., 2012; 132(1): 282—294

[39] Brungart D.S., Durlach N.I., Rabinowitz W.M., Auditory localization of nearby sources. II. Localization of a broadband source, J. Acoust. Soc. Am., 1999; 106 (4), 1956–1968.

[40] Yu G Z, Xie B S, Rao D. Near-field head-related transfer functions of a artificial head and its characteristics. Acta Acousticca (in Chinese), 2012;37(4), 378-385

[41] Møller H, Jensen C B, Hammershøi D, et al. Using a typical human subject for binaural recording, AES 100th Convention, Copenhagen, Denmark, Preprint: 4157; 1996

[42] Wenzel E M, Arruda M, Kistler D J, et al. Localization using nonindividualized head-related transfer functions. J. Acoust. Soc. Am., 1993; 94,111-123

[43] Mackenzie J, Huopaniemi J, Valimaki V, et al. Low-order modeling of head-related transfer functions using balanced model truncation. IEEE Signal Processing Letter, 1997; 4 (2), 39-41

[44] Blommer M A, Wakefield G H. Pole-zero approximations for head-related transfer functions using a logarithmic error criterion. IEEE Trans. on Speech and audio processing, 1997; 5 (3), 278-287

[45] Haneda Y, Makino S, Kaneda Y, et al. Common-acoustical-pole and zero modeling of head-related transfer functions. IEEE Trans. on Speech and Audio Processing, 1999; 7 (2), 188-196

[46] Harma A, Karjalainen M, Savioja L, et al. Frequency-warped signal processing for audio applications. J. Audio Eng. Soc.,2000; 48 (11), 1011-1031

[47] Huopaniemi J, Zacharov N. Objective and subjective evaluation of head-related transfer function filter design. J. Audio. Eng. Soc., 1999; 47(4), 218-239

[48] Jot. J M, Walsh M, Philp A. Binaural Simulation of Complex Acoustic Scenes for Interactive Audio, AES 121st Convention, San Francisco, U.S.A., Preprint: 6950; 2006

[49] Møller H, Hammershøi D, Jensen C B, et al. Transfer characteristics of headphones measured on human ears. J. Audio. Eng. Soc., 1995; 43(4), 203-217

[50] Wightman F L, Kistler D J. Headphone simulation of free-field listening, I: stimulus synthesis. J. Acoust. Soc. Am., 1989; 85 (2), 858-867

[51] Schroeder M R, Atal B S. Computer simulation of sound transmission in rooms, Proceeding of IEEE, 1963; 51(3), 536-537

[52] Bauck J, Cooper D H. Generalization transaural stereo and applications. J.Audio.Eng.Soc.,1996; 44(9), 683-705

[53] Bai M R, Tung C W, Lee C C. Optimal design of loudspeaker arrays for robust cross-talk cancellation using the Taguchi method and the genetic algorithm. J.Acoust.Soc.Am,2005; 117(5), 2802-1813

[54] Kirkeby O, Nelson P A, Hamada H. The "Stereo Dipole"—a virtual source imaging system using two closely spaced loudspeakers, J. Audio Eng. Soc., 1998; 46(5), 387-395

[55] Kirkeby O, Nelson P A, Hamada H. Local sound field reproduction using two closely spaced loudspeakers. J. Acoust. Soc. Am., 1998; 104(4), 1973–1981

[56] Takeuchi T, Nelson P A, Hamada H. Robustness to head misalignment of virtual sound image system. J.Acoust.Soc.Am.,2001; 109(3), 958-971

[57] Ward D B, Elko G W. Effect of loudspeaker position on the robustness of acoustic crosstalk cancellation. IEEE Signal Processing Letters, 1999; 6(5), 106-108

[58] Xie B S, Shi Y, Xie Z W, et al. Virtual reproduction system for 5.1 channel surround sound. Chinese Journal of Acoustics, 2005; 24, 76—88

[59] Gardner W G. Reverberation algorithms, In Applications of Signal Processing to Audio and Acoustics (Edited by Kahrs M, Brandenburg K.), Kluwer Academic Publishers, USA, 1998

[60] Sandvad J. Dynamic aspects of Auditory virtual environments, AES 100th Convention, Copenhaqen, Denmark, Preprint 4226; 1996

[61] Bronkhorst A W. Localization of real and virtual sound sources. J. Acoust. Soc. Am., 1995; 98 (5), 2542-2553

[62] Wenzel E M. Effect of increasing system latency on localization of virtual sounds, AES 16th International Conference: Spatial Sound Reproduction, Rovaniemi, Finland; 1999

[63] Brungart D S, Kordik A J, Simpson B D. Effects of headtracker latency in virtual audio displays. J. Audio Eng. Soc., 2006; 54 (1/2), 32-44

[64] Wenzel E M, Miller D J, Abel J S. Sound Lab: a real-time, software-based system for the Study of Spatial hearing, AES 108 th Convention, Paris, France, Preprint: 5140, 2000

[65] Miller J D, Wenzel E M. Recent developments in SLAB: A software-based system for interactive spatial sound synthesis, Proceedings of the 2002 International Conference on Auditory Display, Kyoto, Japan; 2002

[66] Begault D R, Wenzel E M, Godfroy M, et al., Applying spatial audio to human interfaces: 25 years of NASA experience. AES 40th Conference, Tokyo, Japan; 2010

[67] Saviojia L, Lokki T, Huopaniemi J. Auralization applying the parametric room acoustic modeling technique-The DIVA Auralization system, Proceedings of the 2002 International Conference on Auditory Display, Kyoto, Japan; 2002

[68] Blauert J, Lehnert H, Sahrhage J, et al. An interactive virtual-environment generator for psychoacoustic research I: architecture and implementation. Acta Acustica united with Acustica, 2000; 86 (1), 94–102

[69] Silzle A, Novo P, Strauss H. IKA-SIM: A system to generate auditory virtual environments, AES 116th Convention, Berlin, Germany, Preprint: 6016; 2004

[70] Lentz T, Assenmacher I, Vorländer M, et al. Precise near-to-head acoustics with binaural synthesis, Journal of Virtual Reality and Broadcasting, 2006; 3(2)

[71] Zhang C Y, Xie B S. Platform for virtual auditory environment real time rendering system, ACOUSTICS 2012 HONG KONG Conference and Exhibition; 2012

[72] Langendijk E H A, Bronkhorst A W. Contribution of spectral cues to human sound localization. J. Acoust. Soc. Am., 2002; 112 (4), 1583-1596

[73] Kopco N, Shinn-Cunningham B G. Spatial unmasking of nearby pure-tone targets in a simulated anechoic environment, J. Acoust. Soc. Am., 2003; 114 (5), 2856-2870

[74] Krebber W, Gierlich H W, Genuit K. Auditory virtual environments: basics and applications for interactive simulations, Signal Processing, 2000; 80 (11), 2307-2322

[75] Bronkhorst A W, The cocktail party phenomenon: a review of research on speech intelligibility in multiple-talker conditions, Acta Acustica united with Acustica,2000; 86(1), 117-128.

[76] Begault, D R. Virtual acoustics, aeronautics, and communications. J. Audio Eng. Soc., 1998; 46(6), 520-53

[77] Kleiner M, Dalenback B I, Svensson P. Auralization-an overview. J. Audio Eng. Soc., 1993; 41(11), 861-875

[78] Song W, Ellermeier W, Hald J. Using beamforming and binaural synthesis for the psychoacoustical evaluation of target sources in noise. J. Acoust. Soc. Am., 2008; 123(2),910-924

[79] Toole F E. Binaural record/reproduction systems and their use in psychoacoustic investigation, AES 91st Convention, New York, USA, Preprint:3179; 1991

Chapter 7

Auditory Distance Estimation in an Open Space

Kim Fluitt, Timothy Mermagen and
Tomasz Letowski

Additional information is available at the end of the chapter

http://dx.doi.org/10.5772/56137

1. Introduction

Auditory spatial perception is the ability to perceive relative locations of sound sources in the environment and the spatial character of the surrounding acoustic space. Any property of an auditory event causing a rise to spatial sensation is called a spatial cue. Specific types of judgments resulting from spatial cues are categorized and discussed in the psychoacoustic literature as horizontal localization, vertical localization, auditory distance estimation, and spaciousness assessment. While judgments of directions toward sound sources received considerable interest in psychoacoustic literature, the judgments of auditory distance, and especially the judgments of spaciousness, received much less attention.

Human horizontal and vertical localization judgments and formal and methodological issues related to directional localization of sound sources have been recently reviewed by Letowski and Letowski [1]. Comprehensive summaries of the issues related to auditory distance estimation have been published by several authors including Coleman [2], Blauert [3], and Zahorik et al. [4]. However, these summaries were based on auditory research conducted primarily in closed spaces and at relatively short distances up to about 25 m. Very few studies were reported to be conducted in an open space and they never involved distances exceeding 50 m.

The present chapter is intended to address the distance estimation issues for distant sound sources in an open space and discuss them in the context of our current knowledge of auditory distance estimation. The first part of the chapter provides a comprehensive review of concepts related to auditory distance judgments. It also includes an overview of environmental conditions that effect sound propagation in both closed and open spaces. The second part provides new distance estimation data for free field sound sources located at distances from 25 to 800 meters and uses these data as a basis for a discussion of the environmental variables affecting auditory distance estimation in an open space.

2. Distance perception

Distance perception, sometimes referred to as ranging [5], is the human ability to determine the distance between oneself and a target in space or the distance between two targets in space. The distance to a target can be judged on the basis of its visual, olfactory, and/or auditory properties. Distance judgments may have a form of distance discrimination or distance estimation. Distance discrimination is a relative judgment of the distance in terms of further-closer, less-more, or same-different. Distance discrimination threshold is calculated as a fraction (percentage) of the distance change that is noticeable by the observer. Distance estimation is an absolute judgment about distance in terms of meters, feet, or time to travel; categorical judgment of distance in terms of near-far or predetermined categories; or a direct-action estimation of distance by reaching for the target or walking toward the target. The first two classes of judgments are *explicit estimations* while the third one is an *implicit estimation* (e.g., [6]). Perceived and physical distance seem to be in general monotonically related but can be quite different. In general, human estimation of distance is much less accurate than the determination of angular direction and observers normally underestimate the magnitude of distance.

There are three basic dichotomies that can be used in classifying distance judgments. The first dichotomy divides distance judgments into static (explicit, no-action) and dynamic (implicit, directed-action) behaviors of the judges (observers, listeners). In static (no-action) estimation the judge estimates the distance to a given target from his/her stationary location. These estimates are usually numerical but also can be comparative in relation to other objects in space. In implicit (directed-action) estimation the observer reaches for (e.g., infants) or walks toward (e.g., blindfolded) a target.

The second dichotomy refers to static (stationary) and dynamic (moving) behaviors of the targets. Although dynamic behaviors of judges and target are discussed in the theoretical part of this chapter, the main focus of the chapter is on human ability to assess the distance numerically from a stationary position (explicit estimation). In an open space and for long distances the directed-action (implicit) estimation is often impractical and in many cases unrealistic.

The third dichotomy divides distance judgments into egocentric judgments and exocentric judgments [7]. Egocentric judgments, or body-centered judgments, are the judgments where the point of reference is the observer's location in space. The specific subjective reference point that people use for egocentric visual judgments is the point that lies between the eyes of the observer. In the case of auditory judgments it is the midpoint of the interaural axis of the listener [8-9]. The estimation that the target is located at a certain distance from the observer is an egocentric judgment. In the case of auditory judgments the sound source can be perceived as located either in the head of the listener (such situation takes place in most headphone listening) or outside of the head. In the latter case, the sound source can be in front of, behind, to the left, to the right, above, or below the listener. Exocentric judgments, also called allocentric or geocentric judgments are based on the external frame of reference and are independent of the actual location of the observer. The location of one target in space is referenced to the

http://dx.doi.org/10.5772/56137

location of another target (e.g., a landmark) or to the axes of the external frame of reference. Giving the response as *further north* rather than *further to the right* is an exocentric judgment. This chapter is limited to auditory egocentric judgments and exocentric judgments are not discussed.

3. Auditory distance estimation

Auditory distance estimation is an estimation of a distance to a sound source on the basis of perceived sound. Estimated distance is perceptual measure of a physical distance. The goal of auditory distance estimation is to determine the perceived location of a real or phantom sound source generating a specific auditory event. Such judgments can be made in real surrounding space in respect to natural and electroacoustic (loudspeakers) sound sources or in virtual reality space simulated either through loudspeakers or headphones.

The results of auditory distance judgments are dependent on the availability of several auditory distance cues. Depending on the state of motion of both the sound source and the listener the distance estimation cues are usually classified as static cues (stationary sound source and listener) and dynamic cues (moving sound source or listener) [10-11]. The five basic *static cues* include: sound intensity, direct-to-reverberant energy ratio, sound spectrum, level of background noise, and auditory parallax (interaural differences). The *dynamic cues* include motion parallax and acoustic tau effect (estimated time-to-contact). Please note that static cues operate as well in both the static conditions and the dynamic situations when the either the listener or the sound source is moving. In this chapter only stationary sound source and stationary listener situations are considered.

Another important characteristic of the distance cue is the absolute or relative character of the cue. Absolute cues are those that do not require the listener's familiarity with the sound source and surrounding environment in making distance estimates. Relative cues are those that do. Sound intensity, sound spectrum, and background noise are relative cues and all others are absolute cues. In order to make an informed (relatively accurate) distance judgment using relative cues the listener must be familiar with the sound source (have *a prior* knowledge about sound emission level) and surrounding environment. A prime example of a relative cue is sound intensity. Sound intensity alone is insufficient for the listeners to determine the actual distances to an unfamiliar sound source since its original sound intensity is unknown to the listener [12]. However, with increasing familiarity with both the given sound and surrounding environment distance judgments based on sound intensity can become quite accurate [2, 13].

Other non-specific factors contributing to auditory distance estimates are the listener's expectations, past experience, and non-auditory cues (e.g., visible objects). For example, whispered speech (produced typically at a level of about 30 dB SPL at 1m) is expected by the listener to come from a nearby sound source whereas, normal (conversational) speech (65 dB SPL at 1 m) and a shout (90 dB SPL at 1 m) from much larger distances [14-15]. Therefore, it should be expected that the distance to artificially amplified whispered speech produced by a distant sound source will most likely be greatly underestimated by the listener because a whisper is expected to come from a relatively close distance. More in-depth discussion of acoustic cues and other general factors affecting distance estimation judgments may be found elsewhere (e.g., [4, 16]).

Auditory distance is a prothetic (ratio scale) perceptual continuum. It has the natural zero point (egocenter point) and a unit of measurement (e.g., meter) [17-18]. Each prothetic continuum (y) is exponentially related to the underlying physical dimension (x) by a psychophysical Power Law $y=kx^n$ (see Stevens Power Law [17-19]). In case of distance perception the Power Law has the form

$$PD=kd^{a} \tag{1}$$

where PD is the perceived distance, d is the physical distance, α is the sensitivity of the observer to the perceived distance, and k is a constant dependent on the unit of estimation. If $\alpha=1$, then the changes in the physical or intended distance to the target are accurately perceived. If k=1 and $\alpha<1$ the distance is underestimated and when k=1 and $\alpha>1$ then the distance is overestimated.

In the case of vision, egocentric visual distance estimates are nearly linearly related to physical distances for short distances up to 15-20 m [20-25]. At larger distances observers begin to underestimate the physical distance with estimates converging at a certain asymptotic ceiling (visual horizon) [26-29]. In the case of audition the same general relationship exists but the degree of distance underestimation is greater and the auditory horizon [30] is achieved earlier. The distance to the horizon depends on the listener, available auditory cues, and the acoustic environment, thus it can vary from one situation to another. Zahorik [31] compared results of 10 studies (33 data sets) and reported that the average exponent of the exponential function as $\alpha=0.59$ (SD=0.24) and the constant of proportionality as k=1.66 (SD=0.92). The exponents fitted to individual data ranged from 0.15 to 0.7 and varied much larger between the listeners than between the test conditions (environments). His own study conducted in virtual space (distances from 0.3 m to 14.0 m) resulted in $\alpha=0.39$ (SD=0.13) and k = 1.32 (SD=0.56). In a later study, Zahorik et al. [4] expanded the analysis conducted by Zahorik [31] on the results of 21 studies (84 data sets) and reported the average exponent as $\alpha=0.54$ and the constant of proportionality as k=1.3.

Several studies performed both in real and simulated (headphones) environments indicated that at short physical distances, the perceived distance increases almost linearly with the physical distance [30, 32] or the listeners slightly overestimate its value [4, 33-37]. The tangent of the initial slope of the performance function is close to unity and it can be said that for short distances the auditory distance is approximately a linear function of the physical distance. This range is limited to 1-3 m in both real and virtual environments and it varies depending on both the listening conditions and the listeners [4, 14, 30, 32].

At larger distances (3-48 m) listeners increasingly underestimate the actual distance to a sound source although the distance judgments are slightly more accurate with implicit (e.g., walking toward the source of sound) than explicit (numeric estimation of the distance when both sound

http://dx.doi.org/10.5772/56137

source and the listener remain stationary) estimation (e.g., [37]). In both cases, however, the degree of the underestimation was critically dependent on the availability of specific auditory distance estimation cues, the listener's familiarity with the sound source, visibility of the environment, and the listener's expectations [4, 16, 38-39]. In general, the distance estimates were the most accurate in the case of live talkers [15, 30, 40]. It is also noteworthy that in the case of reproduced speech phrases listeners can make relatively accurate estimates to a source playing natural speech but fail when the speech is played backwards [38, 41].

Regrettably, despite an extensive knowledge accumulated to date about auditory distance perception to sound sources located at short and intermediate distances in enclosed spaces (both anechoic and reverberant) it is still unclear to what extent this knowledge may be applied to sound sources located in an open field at large distances (100 m and more) and operating under various atmospheric conditions. It is unknown what specific role auditory distance cues will have under such conditions and how the open field conditions may affect listener's expectations and perception. A short review of sound behavior under various propagation conditions is provided below.

4. Sound propagation in space

The egocentric auditory distance is the apparent distance from a listener to a sound source. This distance is dependent on the number of auditory cues resulting from the characteristics of the sound source, abilities of the listener, and factors related to sound wave propagation in the surrounding space. The basic sound source, environment, and listener properties that affect auditory distance estimation judgments are shown in Figure 1.

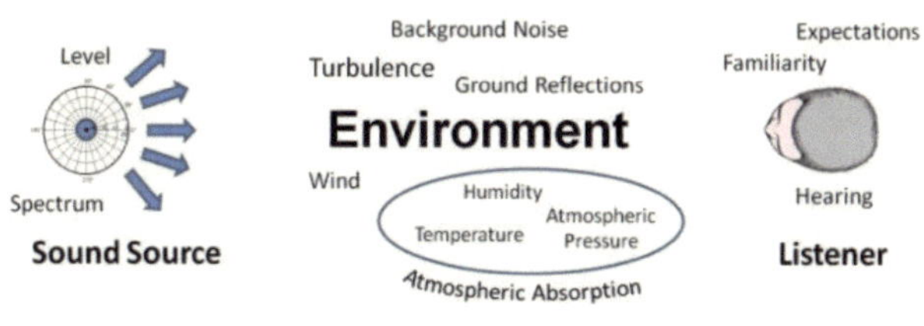

Figure 1. Basic variables that affect auditory distance judgments in an open environment. In a closed environment the additional variables are reflections from space boundaries (echoes and space reverberation) while some environmental variables not present.

4.1. Spherical wave propagation

For an ideal point source (acoustic monopole) radiating sound energy in an unbound sound field (free field), sound energy spreads in all directions (wave front spreading) and the sound intensity I at a given point in space is a function of distance r from the sound source

$$I = \frac{W}{4\pi r^2}, \tag{2}$$

where W is the power of the sound source [watts]. The equation (2) is commonly referred to as the *inverse-square law*. This law applies only to the ideal omnidirectional sound source operating in unlimited space and in the ideal medium, which does not attenuate sound energy. Based on equation (2), the sound intensity level i radiated by the sound source decreases at the rate of 6 dB for every doubling of the distance[1] from the point-like sound source (e.g., idling car) to the observer (listener) according to the formula

$$\Delta i = 10log\frac{I_2}{I_1} = 20log\frac{r_2}{r_1}, \tag{3}$$

where Δi is the difference in the sound intensity level between the sound source location and the observation point and I_1 and I_2 are the sound intensities at the sound source and at the observation point, respectively. Please note that the 6 dB rate of sound decay means that sound intensity decreases four times and sound pressure decreases twice per doubling of the distance. In calculating sound intensity level (dB IL) and sound pressure level (dB SPL) existing at a specific point in space, the common reference values are I_o=10^{-12} W/m^2 and p_o = 10^{-6} Pa, respectively. The 6 dB decay per doubling of the distance only applies to free-sound field or anechoic conditions.. Typical sound decay outdoors over soft ground is about 4.5 dB per doubling the distance. In reverberant environments the decrease is even less, e.g. 4.25 dB in a normal room, due to sound reflections from space boundaries [43].

Assuming that sound intensity at the sound source location is always measured at the distance r_1=1 m, the equation (3) can be reduced to

$$\Delta i = 20\log(r_2). \tag{4}$$

Equations (3) and (4) are valid for an ideal sound source operating in a free sound field but would fail in the presence of reflective surfaces where the sound attenuation with doubling the distance can be expected to be no more than 4-5 dB (e.g., [43]).

Real sound sources, unlike the ideal point source, have finite dimensions and cannot be treated as point sources in their proximity. The sound waves produced by various parts of a real sound source interact in the space close to the source's surface creating; due to constructive and destructive interference of multiple waves originating from the sound source's surface; a complex pattern of spatial maxima and minima of sound intensity. In this region the sound intensity does not obey the inverse-square law and the particle velocity is not in phase with sound pressure. However, at some point in space these separate pressure waves combine together to form a relatively uniform front propagating away from the source. The distance from the sound source where the pattern of spatially distributed maxima and minima merges

1 . In the case of a line sound source, such as moving train or busy highway, producing cylindrical wave, doubling of distance from the sound source results only in a 3 dB reduction of sound intensity level.

in a uniform waveform front is approximately equal to the wavelength (λ) of the radiated sound [43]. The sound field where the sound source can be treated as a point source and the sound wave can be treated as a plane wave is called the *far field*. The area near the sound source where these conditions are not met is called the *near field*.

Most real sound sources are not omnidirectional as the point sound source and radiate most of their energy in certain specific directions. Such sound sources are called directional sources and can be further referred to as dipole, quadrupole, etc. The directionality of a sound source is captured by its *directivity factor* Q and it needs to be taken into account in calculating sound intensity existing at a given distance and direction. Factor Q depends on sound frequency and is equal to one (Q=1) at low frequencies when the wavelength of a sound wave is large in comparison to the dimensions of the sound source and the sound source is effectively omnidirectional. Factor Q can be as large as 10 or more for very directional sound sources. The logarithmic form of the factor Q

$$DI = 10 log Q, \tag{5}$$

is called *directivity index DI* and is expressed in dB.. For an omnidirectional sound source radiating into unlimited free space, DI=0. For the same sound source radiating energy over ideal reflective surface (hemispherical radiation), DI=3 dB [49]. To account for sound source directivity the equation (2) can be modified as

$$I = \frac{QW}{4\pi r^2}, \tag{6}$$

where Q is the directivity factor of the sound source. This equation is only valid for the observation point that is located on the main radiation axis of the sound source.

4.2. Atmospheric attenuation

In a real medium, such as air, sound energy propagating through the medium not only spreads in different directions but is also absorbed by the medium resulting in an exponentially decaying of energy described as the *inverse exponential power law* also called *Beer-Lambert law*. According to this law

$$I = I_o e^{-\alpha d}, \tag{7}$$

where I_o and I are sound intensities at the sound source and the observation point, respectively, d is the distance between these two points, and α is the absorption coefficient of the medium. Absorption of sound energy by a medium, called *atmospheric absorption*, is the result of internal friction within the medium that converts acoustic energy into heat. The basic mechanisms of atmospheric absorption are heat conduction, shear viscosity, and molecular relaxation processes [44]. The amount of energy loss caused by these mechanisms depends on sound frequency, temperature, and atmospheric (static) pressure within the medium and, in case of molecular relaxation processes, on the humidity of the medium (air). This means, that changes in meteorological conditions (weather) have a large effect on sound propagation. Note that although light rain, snow, and fog have relatively very small effects on sound propagation, their presence at larger quantities affects air humidity. The relations between the amount of sound energy absorbed at given frequencies by a medium and meteorological conditions (temperature, atmospheric pressure, and humidity) are complex and non-monotonic functions and the actual amount of resulting absorption depends on specific combinations of these conditions. For example, sound absorption at the temperature of 30 °C is greater for relative humidity of 10% than for 40% while the reverse is true for the temperature of 15 °C (e.g., [45]).

Combining equations (6) and (7) we can predict sound intensity in a real medium as

$$I = \frac{QW}{4\pi r^2} e^{-\alpha d}. \tag{8}$$

At intermediate distances, up to approximately 200-300 m, and at low frequencies the loss of sound energy due to atmospheric absorption by a laminar (not turbulent) medium is usually small (less than 1 dB) and can be neglected for practical purposes [46]. However, at large distances and high frequencies energy loss due to atmospheric absorption can be quite large and exceed the loss caused by a three-dimensional spread of energy. The effect of atmospheric absorption on sounds with high frequency energy above 10 kHz "can become distinctly audible at distances as short as 15 m" (3, p126).

The relationship between the coefficient of absorption (α), sound frequency, and temperature, atmospheric pressure, and relative humidity of the propagating medium can be calculated as

$$\alpha = 8.686 f^2 \sqrt{\tau} \times \left[\frac{1.84 \times 10^{-11}}{\rho} + \frac{(b_1 + b_2)}{\tau^3} \right], \tag{9}$$

where f is sound frequency in Hz, τ is relative temperature ($\tau = T/T_{20}$ in K; T_{20}=293.15 K), ρ is relative atmospheric pressure ($\rho = p/p_n$ in Pa; p_n=101,325 Pa), r_h is relative humidity in %, and b_1 and b_2 are complex coefficients dependent on relative humidity r_h in %, relative temperature τ, sound frequency f, and relaxation frequencies f_n and f_o of nitrogen and oxygen (see ISO 9613-1:1993(E) standard [47], Southerland and Daigle [44], or Salomons [48] for more detailed description of b_1 and b_2 coefficients, which are functions of some of the variables listed above). According to this formula, the coefficient of absorption is proportional to the square of the frequency and is a complex function of weather conditions. The formula is valid for pure tones and narrow-band noises. Its accuracy is estimated to be ± 10% for 153 < T < 323 K, 0.05 < h (concentration of water in the atmosphere; $h = r_h (p/p_n)$) < 5%, p > 200,000 Pa, and 0.0004 < f/p < 10 Hz/Pa [48, p111]. An example of the dependence of the absorption coefficient on frequency for a specific set of environmental conditions is shown in Table 1. Note, however, that equation (9) does not take into account the presence of wind and properties of the ground's surface.

Spherical spread of sound energy (equation 2) and atmospheric absorption (equation 7) are two main sources of attenuation of energy of the propagating sound. However, there are also

http://dx.doi.org/10.5772/56137

	25	50	100	200	400	800	1600	3150	6300
f_c	31.5	63	125	250	500	1000	2000	4000	8000
	40	80	160	315	630	1250	2500	5000	10000
	0.018	0.07	0.25	0.77	1.63	2.88	6.3	18.8	67.0
A	0.028	0.11	0.37	1.02	1.96	3.57	8.8	29.0	105.0
	0.045	0.17	0.55	1.31	2.36	4.58	12.6	43.7	157.0

Table 1. Atmospheric absorption coefficient a (in dB/km) for the preferred 1/3-octave center frequencies f_c (in Hz) [T=283.15 K (10°C); r_h=80%; p=101,325 Pa (1 atm)].

several others. Sound waves propagating close to the ground surface are absorbed and reflected by the ground. This additional factor affecting sound propagation is called ground attenuation. Constructive interactions between direct and reflected sound waves may increase the sound level at the listener up to 6 dB. Destructive interaction may in the worst case completely cancel out the sound. In general, the softer the ground the greater ground attenuation in reference to an ideal reflective surface. The overall amount of ground attenuation depends on the type of ground (ground impedance), sound frequency, the distance over the ground, and the heights of both the sound source and the listener above the ground surface. In the case of a grassy field the ground absorption is most pronounced in 200-600 Hz range and extends toward higher frequencies [44, 49]. The closer the sound source is to the ground surface the greater amount of ground attenuation and greater attenuation of energy at higher frequencies. Fortunately, in many cases ground effects are of little consequence for transmission of sound at heights of more than 1.5 m above ground level [50].

The presence of wind and changes in air temperature with level above the ground surface are additional factors affecting sound propagation. Both these factors are discussed in the next section.

4.3. Wind and other open space effects

When sound travels through still air with uniform atmospheric conditions, it propagates in straight lines. However, wind conditions (velocity and direction), as well as temperature, changes in altitude (height above the ground) affect sound velicity and cause sound waves to propagate along curved lines. Under normal sunny conditions solar radiation heats the earth surface and at lower altitudes the atmosphere is warmer and sound velocity is higher causing a temperature gradient. In the evening, the earth surface cools down and the temperature gradient reverses itself. These two respective temperature conditions are called temperature lapse and temperature inversion. Similarly, wind conditions depend on the height above the ground due to the slowing of the wind at the ground surface due to surface friction. This causes additional wind gradients. When sound velocity decreases with height (upwind sound propagation; daytime sunny warming of the ground) it causes an upward bend of the sound wave (upward refraction). Conversely, when sounds velocity increases with height (downwind sound propagation; evening temperature conversion chilling the ground) it causes a downward bend of sound waves (downward refraction). Upward (downward) refraction of sound caused by the wind can decrease (increase) the expected sound level at the listener location compared to no wind condition by as much as 10 dB depending on the wind strength and change the region of the audibility of sound from smaller or larger.

Atmospheric turbulence, i.e., existence of regions of inhomogeneity in air velocity; caused by local variations in temperature and wind velocity; also affects sound propagation by scattering and focusing sound energy. The changes in sound level caused by atmospheric turbulence can be as large as 15-20 dB, are time dependent, and are characterized by increased sound level in acoustic shadow zones. In addition, all solid objects, such as berms, barriers and towers that are in the path of the propagating sound, disrupt natural propagation of sound energy causing frequency-dependent diffraction and reflection of sound energy. In the case of trees and forests their sound attenuation effect is usually negligible and should only be taken into account at high frequencies (5 dB per 30 m at 4000 Hz [52]). For frequencies above 2 kHz sound attenuation caused by dense forest made of large trees (e.g., jungle) can be estimated as [53]

$$\Delta I\, d = 8.5 + 0.12D, \tag{10}$$

where D is the depth of an infinitely wide belt of forest[2] (m). This estimation is somewhat higher but not much higher than estimation of sound wave attenuation for grassy areas. All these phenomena and mechanisms affect propagation of sound energy in the open space and ultimately affect sound source distance estimations.

4.4. Closed space effects

In closed spaces reflections from space boundaries distort the smooth decrease of sound intensity with the increasing distance from the sound source. Early sound reflections may cause local reinforcement or decrease in sound energy in various locations in the space while the late and multi-boundary reflections fuse together, forming a characteristic delayed trace of sound called reverberation. Reverberant energy is roughly independent[3] of the distance from the sound source and can even dominate overall sound energy at large distance from the sound source. According to the *Hopkins-Stryker Equation* [55] sound intensity at a given point in a closed space is equal to

$$I = W\left(\frac{Q}{4\pi r^2} + \frac{4}{R}\right), \tag{11}$$

where the first element is sound intensity of a direct sound and the second element is sound intensity of the reverberant field caused by space reflections. R is the room constant (in m^2) dependent on total absorption of the space boundaries.

2 This is an empirical formula predicting the amount of sound attenuation (in dB) caused by a certain thickness of a belt of trees. Sound attenuation (in dB) of octave band noises due to sound propagation through dense foliage is given in ISO 9613-2:1966 standard [51].

3 This cannot be said about early reflections, which depend on the position of the sound source in the space.

$$R = \frac{aS}{1 - a}, \tag{12}$$

where S is the total area of room boundaries (m^2) and a is the average sound absorption coefficient of room surfaces. The further from the sound source the smaller contribution of direct sound energy and greater contribution of reverberant energy to the overall acoustic energy in the space. At some distance from the sound source the contributions of direct and reverberant (reflected) acoustic energies are equal and this distance is called *critical distance* $d_{c,}$ which can be calculated from the equation (11) as

$$d_c = 0.141\sqrt{QR}, \tag{13}$$

where V is space volume (m^3), Q is directivity of sound source (dimensionless), and R is room constant expressed in m^2. The relative amounts of direct and reflected energy heard in the room affect listener's perception of the distance to a sound source. Note that in the case of a directional sound source the direct-to-reverberant ratio of sound energy at a given location in the room is also dependent on the orientation of the sound source in respect to room boundaries and the listener's position causing additional dependence of distance judgments on the relative relation of the listener's location to the acoustical axis of the sound source.

5. Distance estimation in an open field

The difficulty of making auditory judgments of distance to a sound source in an open space has been recognized for many years even in relation to relatively short distances [2, 39]. This difficulty dramatically increases in larger spaces and for greater distances. From all the auditory cues discussed above only sound intensity, sound spectrum, and the level of background noise can be used by the listener in a large open field. The only sound reflections available to the listener in an open space are the ground reflections, which are dependent on the form and type of terrain. However, these reflections create a confusing pattern of interferences rather than providing a helpful distance cue to the listener. Still, such an open space is an easier environment for making accurate distance judgments than an urban setting, which is very confusing due to multiple strong reflections coming from unrelated surfaces (e.g., urban canyon).

Meaningful distance estimation to a sound source in a large open space requires the listener to know something about the signal at the source and the types of degradations affecting the signal propagation through the space. This means that the listener needs to be familiar with capabilities of the sound source and be able to predict specific sound source output under given circumstances. In respect to sound propagation through space the sound is degraded by overall attenuation, frequency-dependent attenuation (coloration), reverberation (in woods), and fluctuations in level. In general, it is possible to measure (quantify) each of these kinds of signal changes and even develop a single composite measure of these effects [56] but their effects on auditory distance judgments would be still unknown due to missing field data.

In order to address the existing gap in knowledge regarding auditory distance estimation in an open space we conducted a field study collecting auditory distance estimation data at distances from 25 m to 800 m. To our knowledge this is the first study of this kind and therefore with very limited guidance from literature we had to make several arbitrary decisions regarding the extent of the study and selection of experimental conditions. For example, the study was limited to stationary conditions of both the sound source and the listener, was conducted under relatively stable weather conditions, and only included sound sources located in front of the listener. These specific limitations of the study's design will be evident in the description of the study detailed below. We refer to this study as the *Spesutie Island Study,* in reference to the place where the experimental data were collected.

5.1. Spesutie island study: Method

The Spesutie Island Study was conducted at Spesutie Island, MD on the outdoor test area known as EM Range. The EM Range is an open field approximately 900 m long and 200 m wide. The area is flat, covered with grass, and includes a sand/gravel track encircling the area. Three sides of the area are surrounding by young trees and bushes and the fourth side is separated by additional 50 m of grassy area separating the EM "Range" from a local road. The general view of the area is shown in Figure 2.

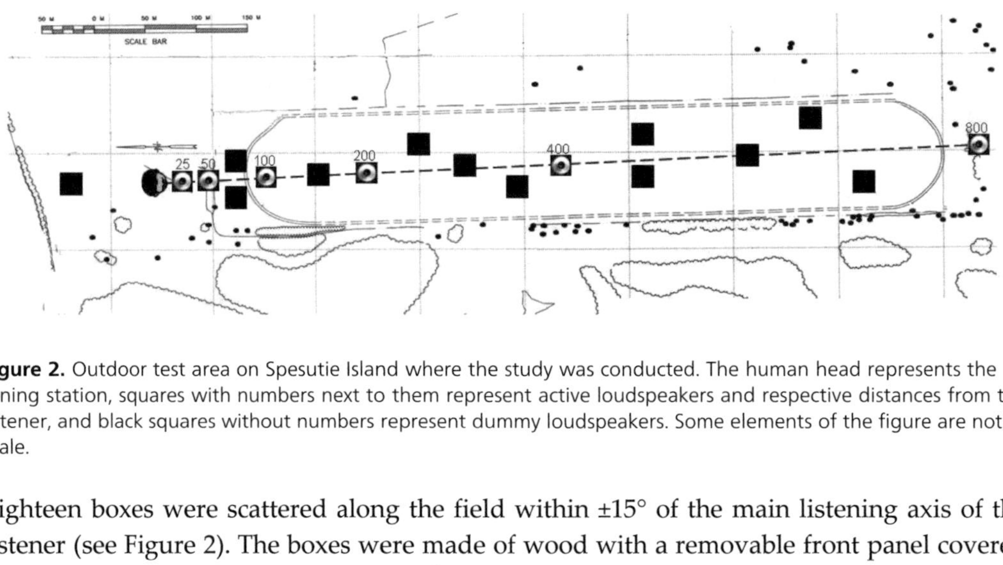

Figure 2. Outdoor test area on Spesutie Island where the study was conducted. The human head represents the listening station, squares with numbers next to them represent active loudspeakers and respective distances from the listener, and black squares without numbers represent dummy loudspeakers. Some elements of the figure are not to scale.

Eighteen boxes were scattered along the field within ±15° of the main listening axis of the listener (see Figure 2). The boxes were made of wood with a removable front panel covered with acoustically transparent cloth. Six of the loudspeaker boxes housed test loudspeakers and other boxes served as decoys. The boxes that contained the test loudspeakers were located at 25, 50, 100, 200, 400, and 800 meters away from the listening station (see Figure 2). The loudspeakers were Electro-Voice Sx500+ stage monitors capable of delivering approximately 120 dB peak SPL at 1 meter distance from the loudspeaker that were fed from Crown 2400 power amplifiers.

The listening station consisted of a table, chair, monitor, keyboard, and a mouse. The station was situated on a concrete slab, protected from sun and bugs by a (2.1m tall) canvas canopy with the walls made of bug netting. The station was also equipped with a Brüel & Kjær 4133 microphone and a Davis Monitor II weather station. The microphone, mounted in an upright position, 1 foot to the left of the listener was used to record actual background noise and test signals during each sound presentation. A weather station, positioned 2 meters to the left of the listener was used to monitor temperature, humidity, wind strength, and wind direction. The data were automatically recorded in the listener file and were used to assess the effects of meteorological variables on sound propagation.

The study was run using a PC desktop computer, TDT System II Signal Processing System, Sony T77 DAT recorder, and supporting hardware and wiring. All equipment not used at the listening station was located in a trailer located at the north end of the range; 50 m to the left of the listening station (not shown in Figure 2). Proprietary software was used to control the experiments and collect listener responses.

A group of 24 listeners between the ages of 18 and 25 participated in the study (M = 21.4; SD = 3.6). All listeners had pure-tone hearing thresholds better than or equal to 20 dB hearing level (HL) at audiometric frequencies from 250 through 8000 Hz (ANSI S3.6-2010 [56]) and no history of otologic pathology. The difference between pure-tone thresholds in both ears was no greater than 10 dB at any test frequency. The listeners had no previous experience in participating in psychophysical studies and were not previously involved in any regular activity requiring distance judgment (e.g., archery, hunting).

Eight natural test sounds were used in the study. Each sound had an overall duration of less than 1s. All sounds, with an exception of generator and rifle shot sounds, which were recorded during another study, were recorded by the authors. The recordings were made with an ACO 7012 microphone and a Sony T77 DAT tape recorder. The respective A-weighted sound pressure levels of the recorded sounds were measured during sound recording. These levels were recalculated for a 1m distance from the sound source and are listed in Table 2. The same sound levels measured at a 1 m distance in front of a loudspeaker were used in the study. The only exception was the *rifle* sound which had a sound pressure level that was too high at a 1m distance to be reproduced and was scaled down by 30 dB to 94 dB A. Spectral and temporal characteristics of all the sounds are shown in Figure 3.

During the study the listener was seated at the listening station and was asked to listen to incoming sounds and respond using a computer keyboard and mouse. An individual test trial consisted of an (1) a warning period indicating the beginning of a new test trial, (2) an observation period and (3) a response period. A yellow-red-green status system light was built into the graphical user interface located on the monitor in front of the listener. The light was used to indicate the warning period (yellow light, 1s), the observation period (red light, 10s), and the response period (green light) when listeners recorded their responses. The length of the response period was not predetermined and listeners could use this time to take short breaks. Listeners were also asked to wait prior to starting the next trial in the presence of occasional extraneous sounds such as an airplane flying over or a car passing by that could interfere with the performed task. To start the next trial, the listener selected the "GO" button on the monitor with the mouse and activated the yellow light which indicated the beginning of the new observation period.

Test Sound	Sound Description	Sound Level
Boltclick	Rifle bolt closure sound	83
Carhorn	Car horn sound	95
Dogbark	Dog bark	88
Generator	Generator sound	74
Joe	Male whisper ("Joe")	72
Rifle	Rifle shot sound	124
Splash	Water splash sound	73
Throat	Throat clearing sound	74

Table 2. List of test sounds and their production levels (in dB A) at 1 meter distance from the sound source.

During each observation period a single test sound or no sound at all was presented. The sound lasted less than 1s and could appear at any time during the observation period. The time when the sound appeared within the observation period was randomized. During the response period the listener was asked (1) to indicate if a sound was present, (2) to identify the presented sound using a 12-item closed-set list of alternatives (which included all the sounds presented in Table 2, plus bird, car engine, airplane, and other), and (3) to determine the distance to the sound source in either meters or yards. No response feedback was given to the listeners but the listeners were told that some sounds may appear very often while others may appear occasionally or not at all. Instructions regarding individual responses and the templates for response input were provided on the computer screen. Prior to the experiment the specific sounds used in the study plus several others listed on the list of alternatives were demonstrated to the listener from a nearby loudspeaker and a short training session was conducted.

One listening block included all seven sounds presented from all six loudspeakers with four repetitions each. In addition 48 blank (no sound) trials were randomly presented in each block resulting in 216 test trials per block. The responses made during the blank trials are not included in the presented data analysis. The order of sounds in each listening block was randomized. Four listening blocks were presented to each listener during a single listening session. The duration of the listening session depended on the duration of the rest periods taken by the listener but was typically 3.0 to 3.5 hours. Large amounts of data were collected during the study but only the auditory distance estimation data collected when the listener correctly recognized the sound are discussed in this chapter. The requirement of correct sound recognition for making distance estimate a valid distance estimation judgment was made to minimize the effects of occasional environmental sounds (birds, cars, remote military sounds, airplanes, etc.) that could have been confused with the test stimuli on listeners' responses.

The study was conducted during a two week period in the month of August. At this time the weather in Maryland is typical of that of the Mid-Atlantic United States. Historically, weather conditions in August in Aberdeen, MD (Spesutie Island area; sea level altitude) are relatively stable with 71% average relative humidity varying from high 50s% (morning) to high 80s% (afternoon); mean temperature during the day in 22-26 °C range (mean 24.1°C) and are characterized by the lowest average wind velocity throughout the year (about 5-6 km/h) [57-58].

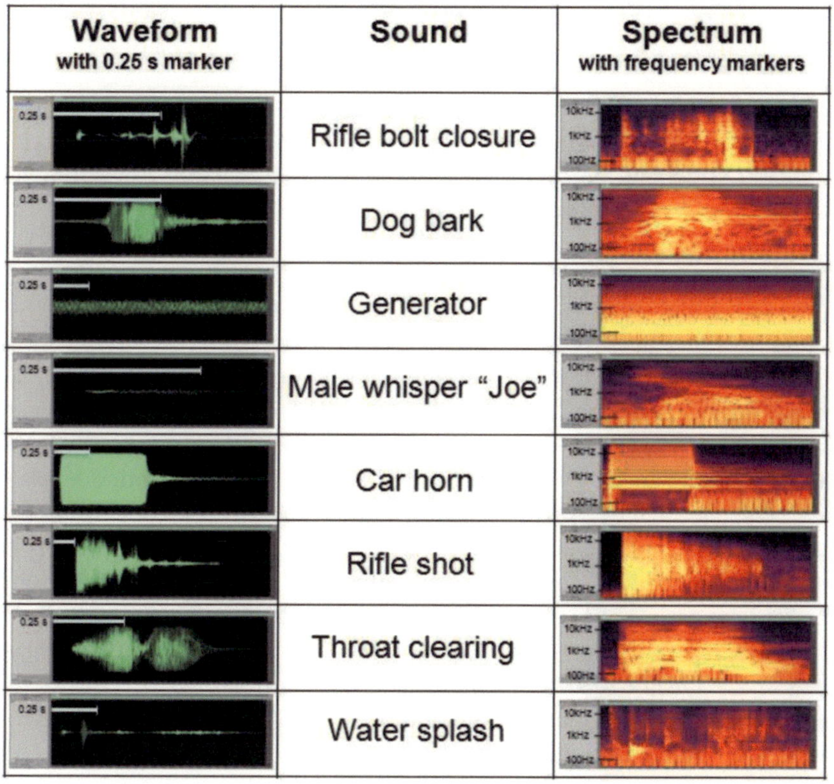

Figure 3. Spectral and temporal characteristics of the sounds used in the study.

The Maryland Department of Natural Resources [59] reports that there are over 400 species of birds and an untold number of insects inhabiting the area surrounding the test site. Sounds made by many of these species created the ambient noise floor that served as a backdrop for our study. The time and temperature of the day also contributed to the acoustic behaviors of some of the wildlife. Many of the insects that contributed to our background sounds were crickets, katydids, cicada, bees, beetles, and grasshoppers. The average weather and noise conditions observed during the study are listed in Table 3. The averages are mean values of the average conditions for individual listening sessions. The overall weather conditions were a bit warmer and drier than average for the area resulting in an average heat index of 31°C.

Parameter	Mean	>Median	Standard Deviation	Unit
Temperature	28.5	29.0	2.3	°C
Relative Humidity	67.6	68.0	0.2	%
Atmospheric Pressure	1.005	1.006	0.018	Atm
Wind Velocity	5.3	4.6	2.3	km/h
Wind Direction	150.0	159.0	37.6	°
Noise Level	50.7	53.0	5.2	dB

Table 3. Mean, median, and standard deviation values of the weather and noise conditions during data collection.

Stronger winds generally came from the South and South-East directions while with many periods of weak wind came from the other directions. The background noise varied between 35-60dB A-weighted depending on the time of the day and weather conditions with a large number of insects producing sounds in the range of 4-8 kHz.

5.2. Spesutie island study: Data

One of the main arbitrary decisions that had to be made in designing the study was the decision about production levels of the loudspeaker-simulated sound sources used in the study. Since the goal of the study was to simulate as much as possible natural sound sources and to learn some basics about the expected distance to an emitting sound source emitting sound in an open space, all recorded sounds were reproduced at their natural recorded levels (except for the rifle shot). This means that each sound was produced at only a single level (see Table 2) by all loudspeakers regardless of the distance of the loudspeaker from the listener. As a consequence, not all the sounds were heard and properly recognized by all listeners when emitted from the distant loudspeakers. The variable audibility of sounds was also exuberated by changes in weather conditions across the study. This was the expected constraint of the implemented study design focused on natural production levels. Obviously, the selected sound events and their levels were selected arbitrarily, but they were representative of specific sound sources and the selected design focused on sound production (as opposed to presentation) level. This design was considered important in an initial study of the effects of sound propagation in an open field on perceived distance to a sound source.

The numbers of valid responses, that is, distance estimations made for correctly detected and recognized sound sources, made by listeners for specific sound source-distance combinations are shown in Table 4. The listeners made close to 100% valid distance estimations for distances up to 100 m and more than 50% valid estimations for distances up to 400 m for all the sounds except for *Joe* and *Throat*. They also made at least 50% valid estimations for *Carhorn* and *Rifle* sounds presented at 800 m distance. The *Joe* and *Throat* sounds were practically inaudible to most listeners beyond 100 m distance. Therefore, in order to avoid making conclusions on the basis of a very limited number of responses for some sound-distance combinations, only the combinations for which more than 50% of responses were collected were generally considered in data analysis. The few exceptions are noted in the text.

http://dx.doi.org/10.5772/56137

Test Sound	Distance (m)					
	25	50	100	200	400	800
Boltclick						
Carhorn						
Dogbark						
Generator						
Joe						
Rifle						
Splash						
Throat						

Table 4. Number of valid responses (detected and recognized sounds) made by the listeners. Black cells: 24-22 responses; gray cells: 18-12 responses; white cells: 10 or fewer responses.

5.2.1. Effects of distance

Distance was the main variable investigated in the study. In order to assess the general effect of distance on auditory distance estimation, estimates made by the listeners for all eight sounds were averaged together for each of the six distances. Two specific cases were considered one, where only distance-sound combinations providing at least 50% of valid responses were considered and two, where all valid responses were averaged together regardless of the actual numbers of responses for specific sound-distance combinations. Both mean and median results of both types of averaging are shown in Figure 4. The standard deviations of the data are not shown since the data are characterized by high variability and standard deviations are in the order of the range of the distance being estimated. Such large variability of the auditory estimation data is normal and is commonly reported (e.g., [4, 12, 60]).

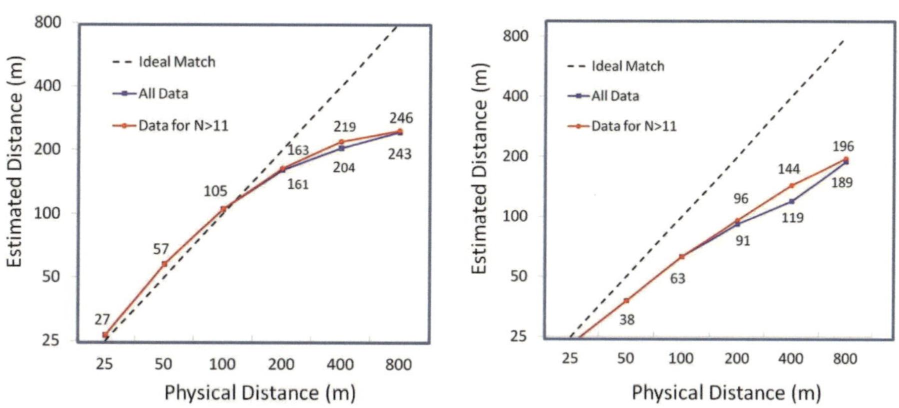

Figure 4. Auditory distance estimation. Mean (left panel) and median (right panel) estimated distance as a function of physical distance for all collected data and for cases where the number of listeners making valid responses was larger or equal to 12. The numbers in the graph are the actual average estimated distances for six physical distances used in the study.

The two curves shown in both panels of Figure 4 are very close to each other despite the quite different number of listeners' responses for 200-800 m data. This supports the general validity of the data collected for sound-distance combinations resulting in 50% or more valid responses. The reported mean curves seem to reach their plateau of about 300 m at the distance of 1000-2000 m that can be hypothesized to be the *auditory horizon* (see [30, 32]) for the listeners in an open grassy field. The shape of the curves agrees with typical curves published in similar studies conducted at close distances and in enclosed environments. They can be approximated by power functions (see equation 1) $PD=12d^{0.41}$ (data for $n\geq12$; $R^2>0.9$) and $PD=12d^{0.46}$ (all data; $R^2>0.9$). The power exponents of both functions are relatively close to the average values reported for shorter distances by Zahorik [31] and Zahorik et al. [4].

The most notable property of the mean curves shown in Figure 4 is that the listeners were either very accurate in their judgments or slightly overestimated the actual distance for distances up to 100 m. Recall that in almost all previous studies conducted in closed spaces such accurate or overestimating judgments were typical for distances not exceeding 1-3 m [32, 34, 61-62, 63]; the last study was conducted in an open space]. Brungart [64] investigated auditory distance estimates over headphones to talkers recorded in open field at distances ranging from 0.25 m to 64 m and reported underestimation of distances larger than 1 m. Visual estimates made in open field at distances at 10 m and beyond are also reported as being underestimated by observers (e.g., [65-66]). These data agree with the general trend in distance estimation judgments described in Section 3. The low intensity sounds coming from larger distances make the differentiation between distances more difficult for listeners. Additionally, listeners tend to expect distant sound sources to be closer than they are in reality due to the typical lack of experience with such judgments and missing cues.

A completely different character of the collected data emerges from the analysis of median values. As shown in Figure 4 (right panel) all distances from 25 m to 800 m have been heavily underestimated by most of the listeners. The observed difference between the mean and median data results from the large variability of the listener responses. The majority of the listeners underestimated all judged distances but several cases of overestimation greatly affected the mean values. Inspection of the data indicated that some listeners had a tendency to overestimate the actual distance to the sound source regardless of the distance and the type of sound source. The latter agrees qualitatively with data reported by Cochran et al. [40] who presented listeners (n=20) with both live and recorded speech stimuli in an outdoor environment at distances from 1 to 29 m. Listeners estimated the distances using magnitude estimation judgment relative to a standard distance and underestimated the longest distance by as much as 30% when the standard distance was close to the listener.

One possible explanation is the fact that some listeners had a tendency to overestimate distances to sound sources across all distances, which may be a sensory influence caused on some listeners by a large visible space and a large number of potential sound source located at large distances. They could expect a greater number of sounds coming from further distances and could react accordingly. Calcagno et al. [67] studied auditory and audio-visual distance estimation in a closed space for distances from 1 m to 6 m and reported that while auditory distance estimates for distances over 2 m underestimated the distance, adding visual cues led to more accurate judgments or even overestimation of distance in the whole range of distances up to 6 m. They hypothesized that auditory distance estimation is affected by visual awareness of the environment, which hypothesis seems to be supported by the estimates made by some of our listeners.

5.2.2. *Effects of sound type*

The distance estimation functions for the individual simulated sound sources used in the study are shown in Figure 5.

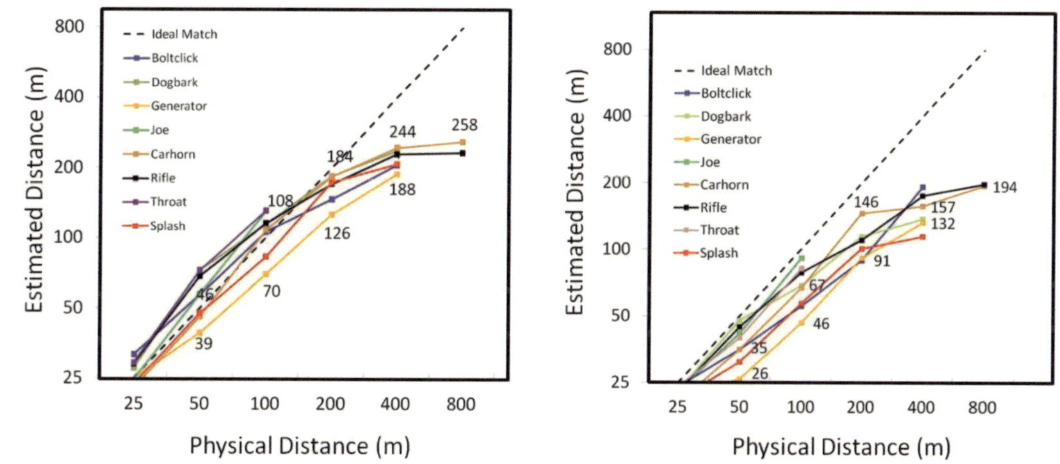

Figure 5. Auditory distance estimation. Mean (left panel) and median (right panel) estimated distance as a function of physical distance for individual sounds and distances where the number of listeners making valid responses was larger or equal 12. The numbers on the graph are the average estimated distances for *carhorn* (top numbers) and *generator* (bottom numbers) sounds.

Inspection of Figure 5 shows that distances to some of the sound sources (*splash, generator*) were underestimated regardless of the actual distance. This can be seen in both mean (Figure 5, left panel) and median (Figure 5, right panel) data representations. In contrast, the distances to sound sources producing relatively low output (*joe, throat*) that could only be heard at short distances were judged accurately (medians) or overestimated by some listeners more than the distances to other sound sources (means). These differences among sound sources may be due to the spectro-temporal properties of emitted sounds, listeners' expectations, or – in the latter case - to a relatively narrow range of effective distances at which these sound were heard. Interestingly, both the *joe* and *throat* sounds differed very much in their both temporal and spectral properties (see Figure 3). Considering this, it seems unlikely that their spectro-temporal properties themselves could be the only or the main factors causing the observed mean overestimation of distances to both of these sound sources. In addition, both sounds are vocal sounds which are familiar to general listeners and should result in fairly accurate judgments. However, due to the requirements of the experimental deign of the study both selected exemplars of sounds were relatively loud for their classes of sounds (whispered *joe* was a voiced whisper). Therefore, it is quite possible that some listeners facing a large open space and hearing louder than expected familiar sounds overestimated the actual distances trying "to use" the whole visually available space. This hypothesis could be verified in the future by conducting a similar study with both sounds presented with different intensities for blindfolded listeners. It may be expected that lack of a visual cue in a form of a large open space could lead to more accurate judgments of both sounds by all the listeners.

The data collected for *boltclick, dogbark, rifle,* and *carhorn* sounds show similar tendency and they mostly influenced the average data discussed in Section 5.2.1. Surprisingly, scaling down the rifle sound by 30 dB was not reflected in distance estimation estimates made by the listeners. This may be attributed to the fact that the actual distance to the "real" rifle location was much beyond the auditory horizon of the listeners. It may also be considered as a finding supporting the theory that the size of visible environment affects (limits, in this case) the range of available distance estimation options (alternatives).

Overall greater underestimation of distances to the *splash* and *generator* sounds was most likely due to expectations and previous life experience of the listeners. The *generator* sound was originally produced by a field generator that could be confused with residential outdoor power equipment, such as a lawn mower, which produces spectrally very similar noise but is typically heard from closer distances. The *splash* sound had the intensity and character typical for this class of sounds but such sounds are seldom heard without close visual effect of splash. A mental image of a visually close event could potentially affected listeners' judgments.

5.2.3. Effects of temperature, humidity, and atmospheric pressure

The two main weather parameters investigated in this study were temperature and relative humidity. Temperature is the measure of the average amount of kinetic energy in the body or environment expressed on a normalized scale. Relative humidity is the ratio of the amount of moisture in the air to the total amount of moisture that can be held at a given temperature, that is, the degree of saturation of air with moisture.

In order to assess the effects of temperature and humidity on auditory distance estimation the data collected during the times of highest and lowest values of both parameters have been analyzed separately. The four extreme weather conditions labeled hot, cool, dry, and humid weather and their temperature and humidity ranges are listed in Table 5. Obviously, they are the extreme conditions in relation to the average weather conditions experienced during the study. Note that temperature and humidity of air are interdependent variables and they could not be absolutely separated for analysis purposes in our study.

Analysis of distance estimation data obtained under the weather conditions listed in Table 6 has been conducted by comparing data collected during pairs of each opposite conditions: hot (5 listeners) and cool (5 listeners) and dry (4 listeners) and humid (4 listeners).

Hot-Cool: The five listeners exposed to the *hot weather* condition performed on the same level as the rest of the listeners. However, the listeners exposed to the *cool weather* condition

Type of Weather	Temperature Range	Relative Humidity Range	Average Temperature	Average Relative Humidity
Hot Weather	29 - 34°C	55 - 75%	31°C	64%
Cool Weather	24 - 27°C	65 - 88%	25°C	78%
Dry Weather	24 - 33°C	50 - 62%	28°C	61%
Humid Weather	24 - 27°C	77 - 98%	26°C	80%

Table 5. Extreme (relative) weather conditions (temperature and relative humidity) recorded during the study.

underestimated the distances for all sound sources more than the rest of the listeners. The mean distance estimations of the *cool weather* group were frequently as much as twice smaller than those of the rest of the group. The behaviors of both groups were very uniform across distances from 25 m to 100 m and they become somewhat random at larger distances where the numbers of responses became quite sparse (all listeners' responses have been included in calculations).

Dry-Humid: For distances from 25 m to 100 m both the *dry weather* and *humid weather* conditions listeners responses differed from the mean values for the whole group. The *dry weather* group provided slightly larger and the *humid weather* group considerably smaller distance estimates than the rest of the group. The behaviors of both groups were the same for all sound sources with one exception. The dry weather condition did not affect the judgments for the *dogbark* sound. For distances above 100 m the effect of the *dry weather* conditions seemed to disappear and above 200 m the effect of the humid weather condition becomes less clear.

Obviously, the above observations need to be treated with caution since they are based on relatively small samples of both the listeners and weather conditions. Since the changes in weather conditions also affect insects' behavior, the weather-related changes in the distance estimates may be affected, and to some degree explained, by the simultaneous changes in the background noise level. These changes are discussed in the forthcoming Section 5.2.5 and additional comments about joint temperature, humidity, and noise conditions are made in that section. In addition, the listeners exposed to the "extreme" weather conditions had their own expectations and experience that could be different from those of others and affected their responses in a unique way.

No effect of atmospheric (barometric) pressure has been noted in the study. Atmospheric pressure is the hydrostatic pressure caused by the weight of air molecules above the measurement point on the Earth's surface. Low atmospheric pressure means that the air is rising and high barometric pressure means that the air is sinking. Atmospheric pressure observed during the study was quite high and relatively stable averaging 1.005 atm and varying from 1.001 atm to 1.009 atm across all listening sessions. Such pressure is typical for very warm weather and was slightly higher than the historically average pressure for the month of August in Maryland. Thus, due to relatively stable pressure conditions during the study no specific effects of atmospheric pressure on distance estimation data were observed.

5.2.4. *Effects of wind*

Wind is one of the major factors affecting sound wave propagation in the environment. Wind effects are quite complex, fast changing (e.g., wind gusts), and confounded by other weather conditions and, as a result, it is hard to assess various wind effects in studies like the current one. Therefore, it was important for the study that all data collection was limited to a relatively stable and weak wind conditions. The average wind speed throughout the study was 5.3km/h (median = 4.6 km/h), with an average direction of 150° (SSE direction). On the Beaufort wind force scale most wind conditions recorded in the study ranged between 0 (calm, less than 1km/h) and 1 (light air, between 1-5.5km/h). There were several (9) sessions with stronger winds ranging from 5.8 km/h to 9.8 km/h but in all cases except one (side wind; no strong perceptual effects) the wind blew downwards (toward the listener). This limited the potential analysis of the wind effects to the comparison between data collected during strong downwind conditions (8 cases) and data collected during no-wind and low-strength-wind conditions (15 cases; 0 to 5.15 km/h; various wind directions) referred later as no wind condition. The results of this analysis are shown in Figure 6.

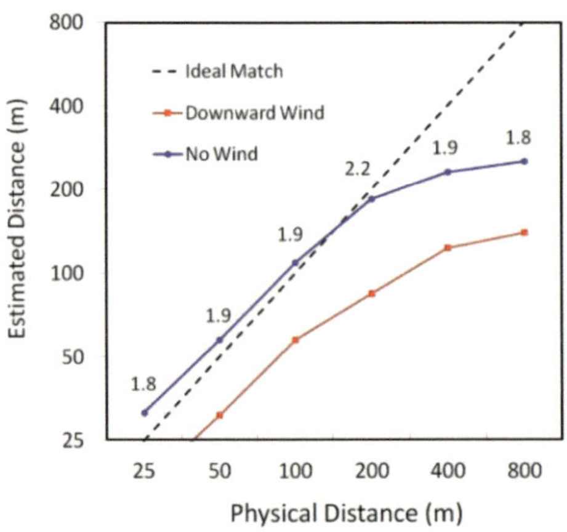

Figure 6. Comparison of auditory distance estimation data for no wind and downwind conditions. The numbers in the graph are the ratios of distance estimates for no wind and downwind conditions.

Under both no wind and downwind conditions the listeners generally underestimated distances to all sound sources. The distance estimates made by the listeners making judgments under no wind condition (M=3.9 km/h; SD=1.0 km/h) were about twice as large as those made by the listeners exposed to strong downwind condition (M=8.2 km/h; SD=1.2 km/h). The results were somewhat dependent on the type of sound with *rifle* (~2.4 ratio) and *carhorn* (~1.7 ratio) sounds being affected the most and the least, respectively. Both of these sounds were the most intense sounds but they greatly differed in spectro-temporal properties. The *rifle* sound was shorter and had lower high frequency content than the *carhorn* sound (see Figure 3). Therefore, it seems that the downward wind enhanced audibility of the *rifle* sound and helped to preserve its less intense high frequency content but such enhancement did not change the perceptual impressions of the listeners in the case of the *carhorn* sound. Recall also that the rifle sound was scaled down by 30 dB during its reproduction.

5.2.5. *Effect of background noise*

The background noise that affected the audibility of sounds produced by loudspeaker-simulated sound sources was for the most part noise produced by ever-present insects. Occasional sounds produced by birds, animals, distant cars, and overflying airplanes were relatively rare, quite distinct, and usually quite short. They could affect one or two of the specific judgments, resulting usually in invalid response, but they did not contribute significantly to the continuous noise present in the field. The average noise level across the study was about 51 dB A and was dependent on the weather conditions and time of the day. Typically, as the day became warmer insect activity decreased making the afternoons quieter than the mornings. As a result most sounds were less audible during cooler mornings than hotter afternoons. The relationship between the noise level and the temperature of air recorded throughout the study is shown in Figure 7.

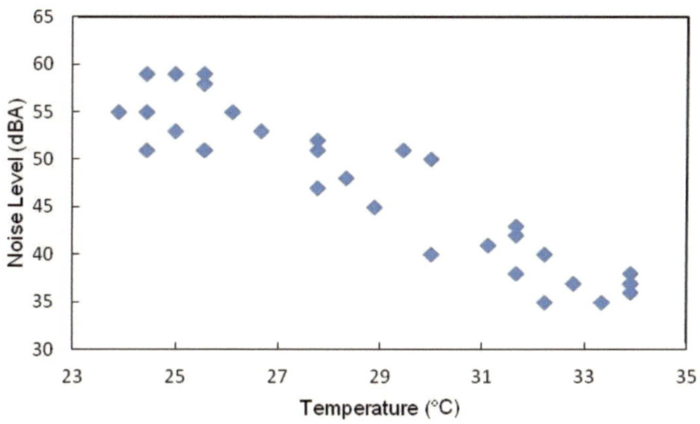

Figure 7. Relationship between background noise level (insects' calls) and temperature of air measured during the study. Not all the points on the graph correspond to actual listening sessions.

The spectral properties of the background noise are shown in Figure 8. The insects' calls were most intense in the frequency band from about 4 kHz to 8 kHz and the noise level resulting from a number of insects' calls decreased by 3-5 dB in the frequency range from ~0.5 kHz to 10 kHz when temperature increased from 28 °C to 32 °C.

As discussed in Section 5.2.3 in general cooler and more humid weather conditions resulted in greater underestimation of the distances to all sound sources. The participants that listened during these weather conditions usually gave closer distance estimates despite the fact that the background noise level under these conditions was higher. However, the negative effect on the audibility of sounds in an open field caused by higher background noise levels made by insects at low temperatures was apparently compensated by the decreasing amount of air

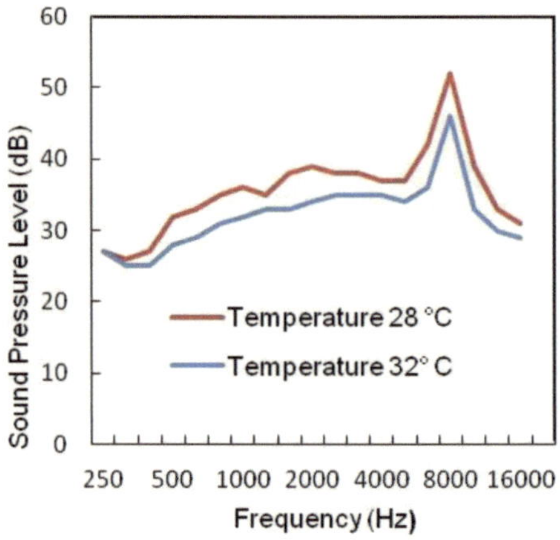

Figure 8. Examples of background noise levels in the morning (28 °C) and in the afternoon (32 °C) of the same day.

absorption (see Section 4.2) caused by increasing humidity and decreasing temperature or by some other factors. Thus, two explanations for the observed effects are possible. First, that the effect of changes in air absorption had stronger impact on the judgments of the listeners than potentially counteracting simultaneous changes in noise level. Second, that poorer audibility of sounds due to higher background noise level was perceptually associated with closer distances to sound sources. The greater the background noise level and the lower signal-to-noise ratio the stronger the listeners' impression that the sound source was relatively near but was masked by background noise. Listeners informally reported that at higher noise levels they "heard" the space as being smaller. Such explanations of the noise effect also agrees with the results of previous research studies conducted in closed spaces regarding the role of *background noise cue*, where higher noise level masked environmental (reverberated) sounds masking the impression that the space was smaller. The discussed effects might also result from specific experience and predispositions of the small number of listeners who were exposed to the "extreme" listening conditions analyzed in our study. Further studies are needed to explain these relationships and answer the related questions.

5.2.6. Individual differences

Distance perception data obtained in the current study are marred by lack of consistency due to listeners' potential lack of ability to use distance estimation cues in open space and large individual differences among the listeners. Typical standard deviations of the group's judgments were close to the size of the physical distance being estimated and quite independent of the type of sound source. The large individual differences and disparities in judgments also have been encountered in closed spaces by other researchers. Recently, Wisniewski at al. [41] used open field recordings reproduced in a closed space and reported substantial individual differences among the listeners in judging auditory distance. The differences ranged from 51% to 77%. However, the listeners made the same general pattern of errors; a finding that is not supported by the results of the present study. Similar, widely varying results

http://dx.doi.org/10.5772/56137

of distance estimation have been reported in visual distance estimation studies conducted in open fields. The results of all these perceptual studies indicate that regardless of sensory input we have not yet found a common relationship between physical and perceived space that is consistent with distance judgments in outdoor contexts [68-69].

6. Summary

The purpose of this chapter was to summarize the state of the art knowledge about the mechanism of auditory distance perception and to report the results of the distance estimation study conducted in an open field for distances in the 25-800 m range. Since this study seems to be the first study of this kind, it actually poses more questions than provides definite answers. A range of listeners' behaviors has been identified but the exploratory nature of this study and the relatively limited number of samples of both the listening conditions and participants advise caution in generalizing the reported data. In addition, interdependence of temperature, humidity, and environmental noise makes some observations tentative that require more rigorous confirmation.

In summary, within the constraints of the reported study, the following conclusions can be made on the basis of collected data:

- Auditory distance estimation judgments in the open field differ greatly among listeners; however, for most listeners the perceived distance and the physical distance are monotonically related.
- The auditory distance judgments in an open field at distances of 25 m and beyond are commonly underestimated compared to the actual distances to sound sources regardless of the distance.
- Some of the listeners participating in the study generally overestimated all distances to the sound sources[4]; this behavior can be explained by either the expectations caused by a large visible space or by lack of an internal concept of auditory distance resulting in the same numeric estimate across a range of physical distances.
- The type of sound source had an effect on the distance judgments; however, some of the observed environmental effects on the perceived sounds were not always clear.
- The effects of temperature, humidity, and environmental noise are interrelated and difficult to separate analytically; however, both higher humidity and lower temperature increased distance underestimation by the listeners in the current study.
- Increased level of environmental noise at lower temperatures affected the audibility of projected sounds but did not seem to affect in a clear way distance estimation judgements.

4 Individual data reported in some previous studies conducted in closed spaces and at shorter distances also indicate that some listeners had a tendency to overestimate most distances.

- Downward wind greatly increased the degree of distance underestimation across all sound sources and distances (upward wind has not been studied).

The authors hope that the results of this study will increase the listeners' awareness of the complex influences affecting listeners' behaviors in an open field under changing weather conditions. However, further studies are needed to expand our knowledge about the nature of auditory distance estimations made under such environmental conditions and to confirm or correct reported findings.

The future studies should include distance judgments in various types of listening environments (such as the in a desert or in the extreme cold), sounds coming from different directions (the back or sides) and a repeated version of the current study with blindfolded participants.

Author details

Kim Fluitt*, Timothy Mermagen and Tomasz Letowski

U.S. Army Research Laboratory (HRED), Aberdeen Proving Ground, USA

References

[1] Letowski, T, & Letowski, S. Auditory spatial perception: Auditory localization. ARL Technical Report ARL-TR-6016. APG (MD): U.S. Army Research Laboratory; (2012).

[2] Coleman, P. D. An analysis of cues to auditory depth perception in free space. Psychological Bulletin (1963). , 60(3), 302-315.

[3] Blauert, J. Spatial Hearing. Cambridge (MA): MIT Press; (2001).

[4] Zahorik, P, Brungart, D, & Bronkhorst, A. W. Auditory distance perception in humans: A summary of past and present research. Acta Acustica (2005). , 91(3), 409-420.

[5] Howard, I. P. Auditory distance perception. In: Howard IP (ed.) Perceiving in Depth, New York: Oxford University Press; (2012). , 3, 277-308.

[6] Servos, P. Distance estimation in the visual and visuomotor systems. Experimental Brain Research (2000). , 130(1), 35-47.

[7] Klatzky, R. L. (1998). Allocentric and egocentric spatial representations: Definitions, distinctions, and interconnections. In: Freksa C, Habel C, Wender KF (eds.) Spatial Cognition- An Interdisciplinary Approach to Representing and Processing Spatial Knowledge. Berlin: Springer-Verlag; 1998. , 1-17.

[8] Cox, P. H. An initial investigation of the auditory egocenter: Evidence for a "cyclopean ear". PhD thesis. North Carolina State University Rayleigh; (1999).

[9] Neelon, F. N, Brungart, D. S, & Simpson, D. (2004). The isoazimuthal perception of sounds across distance: A preliminary investigation into the location of the audio egocenter. Journal of Neuroscience 2004; , 24(35), 7640-7647.

[10] Devallez, D. Auditory perspective: perception, rendering, and applications. PhD thesis. University of Verona Verona; (2009).

[11] Lu, Y-C. h, Cooke, M, & Christensen, H. Active binaural distance estimation for dynamic sources. Proceeding of the InterSpeech-August 2007, Antwerp, Belgium. International Speech Communication Association; (2007). , 2007, 27-31.

[12] Nielsen, S. H. Auditory distance perception in different rooms. Journal of the Audio Engineering Society (1993).

[13] Haustein, B. G. (1969). Hypothesen über die einhorige Entferungswahrnehmung des menschlichen Gehörs (Hypothesis about the perception of distance in human hearing with one ear). Hochfrequensthechnick und Electroakustik 1969; 78 45-57 (in German).

[14] Brungart, D. S. Preliminary model of auditory distance perception for nearby sound sources. In: Greenberg S, Slaney M (eds.) Computational Models of Auditory Function Amsterdam: IOS Press. (2001). , 83-96.

[15] Gardner, M. B. Distance estimation at 0° or apparent 0°-oriented speech signals in anechoic space. Journal of the Acoustical Society of America (1969). , 45(1), 47-53.

[16] Zahorik, P. Auditory display of sound source distance. Proceedings of the 2002 International Conference on Auditory Displays, July 2005, Kyoto, Japan. ICAD; (2002). , 2-5.

[17] Stevens, S. S. The measurement of loudness. Journal of the Acoustical Society of America (1955). , 27(5), 815-829.

[18] Stevens, S. S. On psychophysical law. Psychological Review (1957). , 64(3), 153-181.

[19] Stevens, S. S, & Galanter, E. H. (1957). Ratio scales and category scales for a dozen perceptual continua. Journal of Experimental Psychology 1957;54(6), 377-411.

[20] Fukusima, S. S, & Loomis, J. M. Da Silva JA. Visual perception of egocentric distance as assessed by triangulation. Journal of Experimental Psychology: Human Perception and Performance (1997). , 23(1), 86-100.

[21] Gibson, E. J, Bergman, R, & Purdy, J. The effect of prior training with a scale of distance on absolute and relative judgments of distance over the ground. Journal of Experimental Psychology (1955). , 50(2), 97-105.

[22] Loomis, J. M. Da Silva JA, Fujita N, Fukusima SS. Visual space perception and visually directed action. Journal of Experimental Psychology: Human Perception and Performance (1992).

[23] Rand, K. M, Tarampi, M. R, Creem-regehr, S. H, & Thompson, W. B. (2011). The importance of a visual horizon for distance judgments under severely degraded vision. Perception 2011; , 40(2), 142-154.

[24] Sedgwick, H. Space perception. In: Boff KR, Kaufman L,Thomas JP. (eds.) Handbook of Perception and Human Performance, New York: Wiley and Sons; (1986). , 1, 21.

[25] Gogel, W. C. The analysis of perceived space. In: Masin SC (ed.) Foundations of Perceptual Theory. Amsterdam; Elsevier; (1993). , 113-182.

[26] Loomis, J. M, & Philbeck, J. W. Is the anisotropy of perceived 3-D shape invariant across scale? Perception & Psychophysics (1999). , 61(3), 397-402.

[27] Proffitt, D. R. Distance perception. Current Directions in Psychological Science (2006). , 15(3), 131-135.

[28] Sedgwick, H. Environment-centered representation of spatial layout: Available visual information from texture and perspective. In: Beck J, Hope B, Rosenfeld A. (eds.) Human and Machine Vision. New York: Academic Press; (1983). , 425-458.

[29] 30] Békésy G vonThe moon illusion and similar auditory phenomena. American Journal of Psychology (1949). , 62(4), 540-552.

[30] Zahorik, P. Assessing auditory distance perception using virtual acoustic. Journal of the Acoustical Society of America (2002). , 111(4), 1832-1846.

[31] Bronkhorst, A. W, & Houtgast, T. Auditory distance perception in rooms. Nature (1999). http://www.nature.com/nature/journal/n6719/abs/397517a0.htmlaccessed 16 August 2012).

[32] Ashmead, D. H, Davis, D. L, & Northington, A. Contribution of listeners' approaching motion to auditory distance perception. Journal of Experimental Psychology: Human Perception and Performance (1995). *21*(2), 239-256.

[33] Brungart, D. S, & Scott, K. R. The effects of production and presentation level on the auditory distance perception of speech. Journal of the Acoustical Society of America (2001). , 110(1), 425-440.

[34] Kopco, N, & Shinn-cunningham, B. Effect of stimulus spectrum on distance perception for nearby sources. Journal of the Acoustical Society of America (2011). , 130(3), 1530-1541.

[35] Loomis, J. M. Da Silva JA, Philbeck JW, Fukusima S.S. Perception of location and distance. Current Directions in Psychological Science (1996). , 5(3), 72-77.

[36] Loomis, J. M, Klatzky, R. L, Philbeck, J. W, & Golledge, R. G. Assessing auditory distance

[37] perception using perceptually directed actionPerception & Psychophysics (1998). , 60(6), 966-980.

[38] Mcgregor, P. K, Horn, A. G, & Todd, M. A. Are familiar sounds ranged more accurately? Perceptual and Motor Skills (1985). Pt2) 1082.

[39] Coleman, P. D. Failure to localize the sound distance of an unfamiliar sound. Journal of the Acoustical Society of America (1962). , 34(3), 345-346.

[40] Cochran, P, Throop, J, & Simpson, W. Estimation of distance of a sound source. American Journal of Psychology (1968). , 81(2), 198-206.

[41] Wisniewski, M. G, Mercado, E, Gramann, K, & Makeig, S. Familiarity with speech affects cortical processing in auditory distance cues and increases acuity. PLoS One (2012). http://www.plosone.org/article/info%3Adoi%2F10.1371%2Fjournal.pone.0041025accessed 30 September 2012)., 7(7), 1-8.

[42] Zahorik, P, & Wightman, F. L. Loudness constancy with varying sound source distance. Nature Neuroscience (2001).

[43] Morse, P. M, & Ingård, K. U. Theoretical Acoustics. New York: McGraw-Hill; (1968).

[44] Sutherland, L. C, & Daigle, G. A. Atmospheric sound propagation. In: Crocker MJ. (ed.) Encyclopedia of Acoustics. New York: Wiley and Sons; (1997). , 341-365.

[45] Harris, C. Absorption of sound in air versus humidity and temperature. Journal of the Acoustical Society of America (1966). *40*(1), 148-159.

[46] Albert, D. G. Past research on sound propagation through forest. Report ERDC/CRREL-TR-Hannover (NH): U.S. Army Corps of Engineers; (2004). , 04-18.

[47] International Standards OrganisationAcoustics- Attenuation of sound during propagation outdoors- Part 1: Calculation of the absorption of sound by the atmosphere. ISO Geneva: ISO; (1993). , 9613-1.

[48] Salomons, E. M. Computational Atmospheric Acoustics. Boston: Kluwer Publishers; (2001).

[49] Lamancusa, J. S. Outdoor sound propagation. Noise Control. ME 458: Engineering Noise Control. State College (PA): Penn State University. (2000). http://www.mne.psu.edu/lamancusa/me458/10_osp.pdfaccessed on 9 September 2012)., 10.

[50] Naguib, M, & Wiley, R. H. Estimating the distance to a source of sound: Mechanisms and adaptations for long-range communication. Animal Behavior (2001). , 62(5), 825-837.

[51] International Standards OrganisationAcoustics- Attenuation of sound during propagation outdoors- Part 2: General method of calculation. ISO Geneva: ISO; (1996). , 9613-2.

[52] Aylor, D. Noise reduction by vegetation and ground. Journal of the Acoustical Society of America (1972). , 51(1), 197-205.

[53] Bullen, R, & Fricke, F. Sound-propagation through vegetation. Journal of Sound and Vibration (1982). , 80(1), 11-23.

[54] Hopkins, H. F, & Stryker, N. R. A proposed loudness-efficient rating for loudspeakers and the determination of system power requirements for enclosures. Proceedings of the Institute of Radio Engineers (1948). , 36(3), 315-355.

[55] Brown, T. J, & Handford, P. Sound design for vocalizations: Quality in the woods, consistency in the fields. Condor (2000). , 102-81.

[56] American National Standards Institute ((2010). Specifications for audiometers (ANSI S3.6-2010). New York: ANSI; 2010.

[57] Current ResultsMonthly humidity averages for Maryland. http://www.currentresults.com/Weather/Maryland/humidity-by-month.phpaccessed on 14 September (2012).

[58] Current ResultsAverage temperatures for Maryland in August. http://www.currentresults.com/Weather/Maryland/temperature-august.phpaccessed on 14 September (2012).

[59] Maryland Department of Natural ResourcesMaryland plants and wildlife. http://www.dnr.state.md.us/wildlife/Plabts_Wildlife/index.aspaccessed on 14 September (2012).

[60] Mershon, D. H, & King, L. E. Intensity and reverberation as factors in auditory perception of egocentric distance. Perception & Psychophysics (1975). , 18(6), 409-415.

[61] Kim, H-Y. Relation between sound pressure level and auditory distance perception in anechoic room. *Journal of the Korean Academia-Industrial Cooperation Society*, (2009). in Korean).

[62] Mcmurtry, P. L, & Mershon, D. H. Auditory distance judgments in noise, with and without hearing protection. Proceedings of the 29th Human Factory and Ergonomics Society Annual Meeting, September- 3 October (1985). Baltimore, USA. HFES; 1985, 811-813.

[63] Speigle, J, & Loomis, J. M. Auditory distance perception by translating observers. Proceeding of the IEEE Symposium on Research Frontiers in Virtual Reality, October 1993, Washington (DC), USA. IEEE; (1993). , 92-99.

[64] Brungart, D. S. Speech-based distance cueing in virtual auditory displays. Proceedings of the 44th Human Factors and Ergonomics Society Annual Meeting, 29 July- 4 August (2000). San Diego, USA. Proceedings HFES 2000; , 44(22), 714-717.

[65] Andre, J, & Rogers, S. Using verbal and blind-walking estimates to investigate the two visual system hypotheses. Perception & Psychophysics (2006). , 68(3), 353-361.

[66] Strauss, M, & Carnahan, J. Distance estimation error in roadway setting. The Police Journal (2009). , 82(3), 247-264.

http://dx.doi.org/10.5772/56137

[67] Calcagno, E. R, Abregú, E. L, Eguia, M. C, & Vergara, R. The role of vision in auditory distance perception. Perception, (2012). , 41(2), 175-192.

[68] Jackson, R. E. Individual differences in distance perception. Proceedings of the Royal Society: Biology (2009). , 1665-1669.

[69] Norman, J. F, Crabtree, C. E, Clayton, A. M, & Norman, H. F. The perception of distance and spatial relationships in natural outdoor environments. Perception (2005). , 34(11), 1315-1324.

Chapter 8

Contribution of Precisely Apparent Source Width to Auditory Spaciousness

Chiung Yao Chen

Additional information is available at the end of the chapter

http://dx.doi.org/10.5772/56616

1. Introduction

It has been shown that the apparent source width (ASW) for one-third-octave band pass noises signal offers a satisfactory explanation for functions of the inter-aural cross-correlation (IACC) and W_{IACC}, which is defined as the time interval of the inter-aural cross-correlation function within ten percent of the maximum (Sato and Ando, [18]). In this chapter, the binaural criteria of spatial impression in halls will be investigated by comparing with ASW for the auditory purpose assistant to visual attention, which is called source localization. It was proposed that the ASW could properly define directional impression corresponding to the inter-aural time delay (τ_{IACC}) perceived when listening to sound with a sharp peak in the inter-aural cross-correlation function (ICF) with a small value of W_{IACC}. We supposed that the ASW would be sensed not only with regard to the relative amplitudes between reflections in a hall, but the total arrived energies at two ears through the A-weighting network in the brain, termed as listening level (LL) and the temporal characteristics of sound sources. This hypothesis is based on the fact that the spatial experience in a room will be varied by changing the center frequency of one-third-octave band pass noise signal, and the ASW decreases as the main frequency goes up. For the purpose of this chapter, we shall discuss the relationship among some factors, the geometric mean of sound energies at two ears, the reverberation, IACC, τ_{IACC}, and W_{IACC}, and whether they are independently related to the sound source on a horizontal plane. Finally, we have discussed that the ASW impression varied in accordance with the acoustic characteristics of sound intelligibility.

2. Effects of reverberation time and sound source characteristics to auditory localization

2.1. Physical properties of source signals regarding sound localization in a hall

According to the reports by Morimoto [1] regarding the influences of sound localization of spatial perception in a hall, the reverberation energy (RT60 = 0.3, 0.9 s) may be treated as the first reflection energy (delay time = 80, 160ms). However, the selection of music is exclusively limited to using Wolfgang Amadeus Mozart's Symphony No. 41, Movement IV as a music source. We intended to prove that the sensitivities on the spatial impression of sound localization will vary depending on the structural characteristics of music. Therefore, the other three sound sources: Motif A (Royal Pavane by Gibbon, τe = 127 ms), Motif B (Sinfonietta, Opus 48; IV movement; Allegro con brio by Arnold, τe = 35 ms) and Speech (female, τe = 23ms) were adopted. According to the sound field design theory described by Ando [2], the determining factor of an ideal reverberation time length lies in the effective delay of autocorrelation function (τe) of sound sources illustrated in Figure 1. The reverberation time of our experiments was set at: short (0.3 s), medium (0.9 s) and long (2.0 s) respectively. The judgments of the apparent sound localization were responded from 12 participants by way of scaling using a normal distribution between two horizontal stimuli angles. The primary analyses of correlations between sound source and auditory localization will presumably the different τe proposed by Ando [2]; namely, the significant difference sensation of reverberate image between Motifs will have an influence on human auditory spatial perception of sound sources.

2.2. Analyses of source signals in a hall

The experiences of visual interaction with the direction of sound source at the stage of opera or a classical orchestra have sometimes failed to catch the scene of the performance with respect to the distance or width of the stage. However, it is important and cheering for the audiences to trace and immediately respond to the present player on the stage as if the source directional sensitivity in a diffusing sound field were accurately installed. In this paper, we have tried to compare the source directional sensitivity of spaciousness as caused by early reflections with different azimuth angles. Morimoto [1] reported that of early reflections at the point of subjective equality (it was termed PSE) of spaciousness shows that they are comparable, but early reflection levels seem to be generally slightly lower than the reverberation. That is, the reverberation level correlated well with the early reflections level at the PSE. This means that both energies are fairly proportional to each other and that the average difference is 1.27dB. Barron and Marshall [3] described that the value of lateral energy fraction, as calculated for a series of reflection sequences for two rectangular halls gave virtually identical values no matter whether 80 ms or 100 ms was used as the limiting delay value for the early lateral reflections. Inoue et al. [4] recently reported that the preference of sound impression did not increase with spaciousness throughout, but may have a maximum value at certain spaciousness, that is, the audience does not prefer excessive spaciousness. Hasegawa et al. [5] reported the sound image width was perceived as narrower or wider than the actual presentation region when the sound source width was decreased or increased, respectively by using two loudspeakers were semicircularly arranged. Ando [2] reported the most preferred delay time of early reflections after the direct sound differs greatly between the two Motifs. It is found that this corresponds to effective durations (τe) of the autocorrelation function (ACF) of source music of 127 ms in Motif A and 35 ms in Motif B. To obtain a degree of similar repetitive features of the sound signals, τe values of ACF were analyzed as a phenomenon of stationary random processing (SRP) strictly defined with an infinite length observation (Marple [6]). Concerning SRP for music signal, the estimation of finite length data (2 s) will only obtain an estimation of ACF as Equation (1). As $\tau \ll N$, the estimation of ACF are almost equal to the ACF only in an initial range. Thus, a linear sum of music shows an initial decline of envelope of ACF, and it can be fit to a straight line regression of the power of the normalized ACF (Figure 1). The τe values of ACF of music is defined as it crosses to -10 dB to that of delay.

$$\Phi(\tau) = \frac{1}{N}\sum_{n=1}^{N} x_N(n)x_N(n+\tau) \tag{1}$$

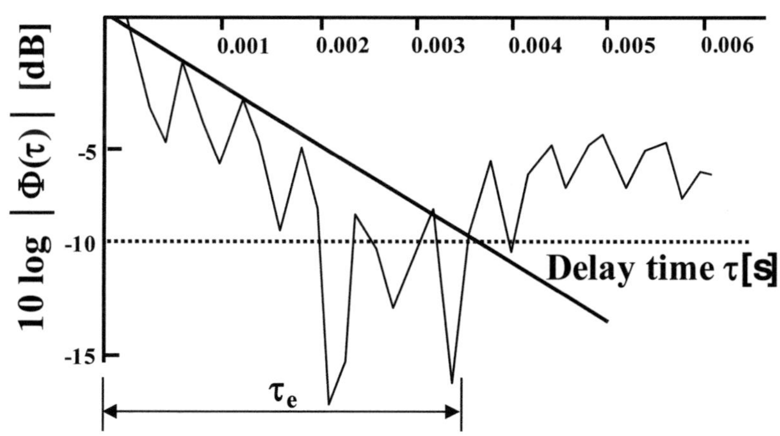

Figure 1. Definition of the effective durations (τe) of the autocorrelation function

In order to represent the geometrical size of a similar room, the delay time of subsequence reflection is introduced as $\Delta t_2 = \Delta t_1 + 0.8\,\Delta t_1$. In this study, the term "auditory localization" was defined as the detection of sound image edge perceived by the auditory event using two loudspeakers as Hasegawa et al. [5].

2.3. Subjective judgments of sound localization

A method of adjustment using LED unit by the subject was employed in this experiment. The subjects could switch the edge direction carefully with a LED unit equipment (Figure 2), as they were asked to answer the angle of edge direction to the maximum possible under the auditory spaciousness they perceived.

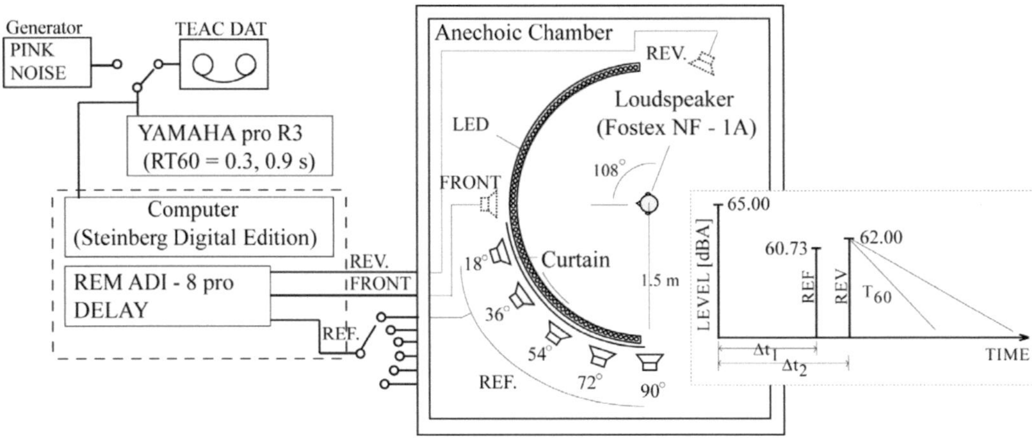

Figure 2. Measuring set-up

1. Apparatus

Figure 2 shows the experimental arrangement. Seven loudspeakers were arranged in the semi-anechoic chamber of the acoustical studio at the Chaoyang University of Technology. The first loudspeaker was in front of the subject at a distance of 1.5m. This 1oudspeaker was used to radiate the direct sound. One further loudspeaker stood at azimuths of +108°, also at a 1.5m distance, used to radiate reverberation. The direct sound was played by digital system controlled on desktop PC derived from a DAT tape recorder (TEAC R-9) and delivered directly to the front loudspeaker. The single early reflection and the reverberant signal with time delay of preferred gap were listed in Table 1. The reverberation time (RT60) was created by a digital reverberator (YAMAHA Pro R3). They were directly delivered to the left horizontal plane by loudspeakers (-18°, -36°, -54°, -72°, -90°) and to the right plane (+108°). Mehrgardt and Mellert [7] measured the transfer functions of the ear canal using the impulse response technique from ten directions of the symmetry plane in a free sound field. The peaks of these functions yield about 8% of the different amounts of the shifted curves at these ten directions from 0° to 180°. The curves of 20 subjects overlap closely, if they are shifted along the logarithmic frequency scale. The angles of the early reflection are in five directions of the frontal symmetry horizontal plane (Figure 2). We could simulate five kinds of sound fields, which all consisted of the direct sound plus reverberation and plus early reflection with arbitrary five azimuth angles. The levels of the early reflections and the reverberant signals relative to the direct sounds which were measured by a noise meter (ONO SOKKEI LA-5110) placed above the head of the subject. For the level measurements (SLOW, A weighting, peak), pink noise was used as a source signal. The LED unit could display each 3.0° azimuth angle; the results of these experiments were scaled using normal distribution function as below, the score was 100 as the answer is absolutely right to the present angle, and 0 showed that the answer was a different angle to the present one.

http://dx.doi.org/10.5772/56616

Motif	**Δt₁**	**Δt₂**	**τe**	*Tempo*
A	127 ms	229 ms	127 ms	slowly
B	35 ms	63 ms	335 ms	quickly
S	23 ms	41 ms	227 ms	quickly

Table 1. Experimental arrangements for the three Motifs

Figure 2 simultaneously shows that the level and time delay structure of each signal was constantly arranged for three Motifs respectively for all situations in our experiments. All the data for three Motifs are shown in Table 1.

2. Musical Motif and Subjects

The Motifs used for the experiments were all initial 5s section of Symphony music; they are: (A). Royal Pavane composed by Orlando Gibbons, (B). Sinfonitetta, Opus 48, IV movement composed by Malcolm Arnold, and (S). Speech "In language infuse the T many words become read the small set later." Poem read by a female, recorded by Burd [8] in the anechoic chamber of BBC. Twelve experienced males, ages 25 ± 2 years, with normal hearing sensitivity served as subjects.

3. Procedures

The subject could switch at will between five azimuth angles using LED unit equipment. After each angle adjustment, the experimenter recorded the results from the LED unit to calculate the score with Equation (2). Reverberation times RT60 of 0.3, 0.9 and 2.0s, and the source signal Motif A, B and S were used for the experimental sound field. The early reflection was radiated at different azimuth angles of -18°, -36°, -54°, -72°, and -90° throughout the three Motifs. Each measurement was repeated three times, yielding a total of 135 experimental results altogether for each subject.

$$SCORE = \frac{1}{\sqrt{2\pi}} e^{-\frac{(angle)^2}{2}} \tag{2}$$

2.4. Analyses of perception on source localization

All data for the twelve subjects are shown together in Figure 3. A three-way (Motif * RT60*Angle) factor analysis of variance (ANOVA) indicates significant individual difference between three Motifs and five angles($p < 0.001$, $p < 0.001$) for all experimental conditions. However, the three-way factor analysis of variance indicates less significant difference ($p = 0.029$) between three conditions of RT60. In addition, there is no interference between the three factors for all experimental conditions. This means that all test sound fields could make the subjects perceive spaciousness after the direct sound field no matter what the reverberation time was in the situation of 0.3, 0.9 or 2.0 s. Therefore, the averaged tendency is obvious for three Motifs are obviously higher ($p < 0.001$) as τe of ACF of the source signal is longer itself (Figure 4).

Especially, in the case of angle = -54°, scores are quite consistent; the Motifs are clearly independent with the reverberation time. In the case of angle = -36°, the scores were least since subjective diffuseness could be most intense, the source width image was blurred. We conducted a further observation on the measurements of inter-aural cross-correlation coefficient measured by Ando [2] for three Motifs. The measured values of the magnitude of ICF (IACC) for five azimuth angles from -18° to -90° of early reflections are shown in Figure 5. The results of measurements of IACC measured at both ears for music. Especially for Motif A and B, they are noteworthy in connection with the results of source localization in this study.

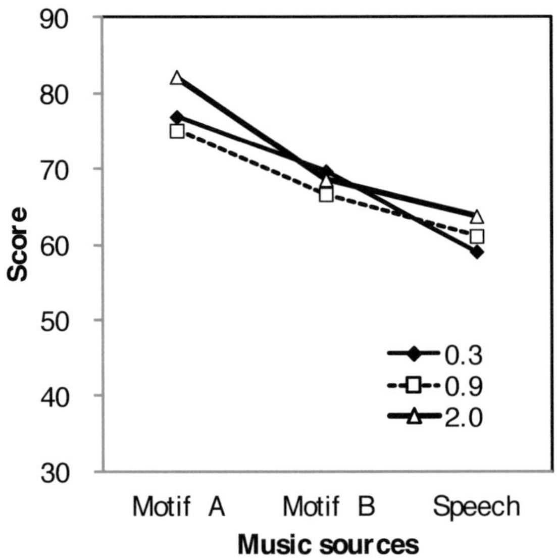

Figure 3. Scores of auditory source directional sensitivity were obtained by changing the coming azimuth angle of early reflection for the three Motifs and different reverberation times. The tendency shows that Motif A obtained the highest accuracy level while speech hit the lowest ($p < 0.01$).

3. Relationship between the envelope of sound image and source characteristics in median plane localization

3.1. Pysical properties of apparent source width regarding sound incident angles

To design an indoor sound field, Ando [9] proposed there are three temporal components involved. They are direct sound, first (initial) reflection and subsequent reverberation. This section was further compared with the spatial perception of a media plane in attempt to detect the edge of the sound envelopment composed by such three components. The relationship between source temporal characteristics and apparent source width (ASW) of spatial impression found in above section were reconfirmed, too. The experiment was arranged the direct sound located in front of the subject ($\eta = 0°$, $\xi = 0°$), and the first reflection came from different

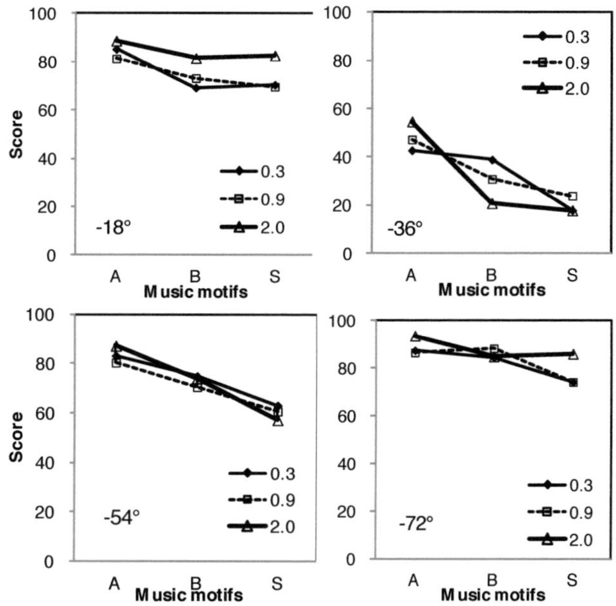

Figure 4. The scores of source width's detection sensitivity function as effective delay of ACF of source in several angles (-18°, -36°, -54°, -72°).

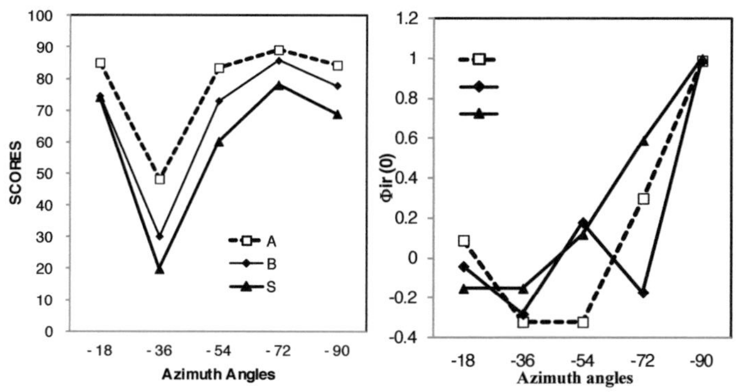

Figure 5. Source directional detection (Left) functions similarly as the tendency of measurements of cross- correlation ($\phi_{lr}(0)$) (Right) for five azimuth angles from -18° to -90° (contra- clockwise) of early reflections.

vertical angles (η = 18°, 36°, 54°, 72°, 90°); and reverberation came with energy at a fixed angle (ξ = 90°). The subjects were instructed to judge the angles of sound image outline in the sound field by keeping attention on some 5 s duration dry sources of the parts of classic music. The purpose of these arrangements is to confirm that whether subjective judgment of image boundary is affected by reverberation time or not. Secondly, is the ability of edge localization independent with the angles of first reflection in media plain?

3.2. Studies reviews of apparent source width at the median plane

We have experienced in edge detection of the sound image envelope in relation to the localization of sound sources on a horizontal plane in an indoor sound field (Chen [10]). According to several reports by Morimoto ([11, 12] and [13]),they confirm that the localization accuracies almost always depend on the presence of spectral cues of median-plane localization, and that most sound images are recognized by both binaural disparity cues and spectral cues at a certain biased direction. However, Morimoto applied only white-noise through a band-pass filter as a sound source, but not a contribution to the aid of building acoustic design. We referred to the results as Morimoto reported [14] on the energy setup of whole reflections within a horizontal plane for apparent source width (ASW) in a hall, and found that source temporal cues have a strong influence on the edge detection of the sound image envelope using the auto- correlation technology proposed by Ando [9]. The purpose of this study focused on the problem of whether or not the localization tests of source images in the upper hemisphere in a median-plane need both binaural cross-correlation cues and dynamically temporal cues. Temporal cues mean that the spaciousness of a sound field depends upon not only on inter-aural cross–correlation but source characteristics themselves. After all, the coming orientations of initial reflections to the audience in a hall indicate an important design theory which is to be improved by source image creation.

Barron and Marshall [15] identified the arrival time of reflections by 80-100 ms after the direct sound. In terms of Morimoto et al. [16], spatial impression comprises of at least the following two components. One is an auditory source width (ASW) which is defined as the width of the sound image fused temporally and spatially with a direct sound's image and the other is listener envelopment (LEV) which is the degree of the fullness of sound images around the listener, excluding the sound image composing ASW. The auditory spaciousness was inquired under initial reflection and reverberation in a concert hall by Morimoto et al. [16]. The difference limen applied to subjective auditory perception. The sound pressure of direct sound as the standard made that of initial and reverberation noticeable. The point of subjective equality (PSE) applied to identify the least sound pressure level under the timing of just-noticeable difference of direct sound energy. The outcomes show that the listener's auditory spaciousness is not affected by delayed reflections and reverberation time at the sound pressure level (SPL) by 1.27 dB between the two reflections.

Room shape, reverberation time and first delay time are often taken into account in designing an indoor sound field; therein, the sidewall planning influential to reflections is valued in particular. However, the azimuth reflection is overlooked. From the reports of [10, 18], there is a correlation between the apparent source width (ASW) and the direct sound, initial reflection and subsequent reverberation of Motifs of which a sound field comprised might compose varied spaciousness of apparent sound source or edge detection of sound image envelopment. The experiments were conducted after validating and verifying the accuracy of the temporal and spatial components to prevent the spatial split. By Chen [10], the temporal characteristics of music do affect the auditory spaciousness of apparent sound source whereas how reverberation time impact on spaciousness is in need of further verification. The human auditory system is sensitive to sounds at frequencies between 1000-4000 Hz pursuant to an

http://dx.doi.org/10.5772/56616

equal loudness contour. Asahi and Matsuoka [17] failed to explain how human ears discern the frequencies. Morimoto et al. [13] employed white noise as the binaural stimuli by 4800 Hz since the azimuth localization depends on the high-frequency sound source in contrast to the low-frequency one. However, the author finds such statement in need of more verification.

This focus of the study is whether or not the localization tests of the source image in the upper hemisphere (Figure 6) in a median-plane need both binaural cross-correlation cues and dynamically temporal cues. Temporal cues mean that the spaciousness of a sound field depends upon not only inter-aural cross-correlation but source the characteristics themselves.

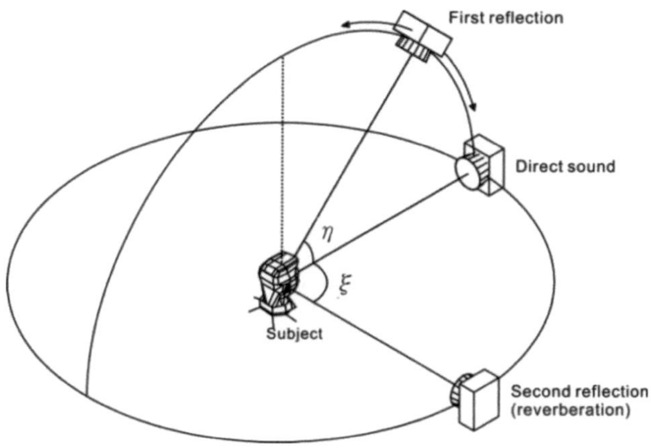

Figure 6. Demonstration of a sound field

3.3. Subjective judgments of source envelope at the median plane

Figure 7 shows how the subject perceived the sound. There were direct sounds in front of the subject (ξ=0°) with first reflection at vertical angles (η = 18°, 36°, 54°, 72°, 90°) and second reflection (reverberation) in front of the subject at 90° (ξ= 90°).

1. Arrangement

The spaciousness consisted of the three components which involved direct sound, initial reflection and reverberation and was surveyed to identify the degree of edge detection on sound envelopment in the upper hemisphere in a median-plane excluding other unwanted factors. First, the subject reported that the perceived angle seated at a specified chair of a semi-anechoic chamber by a semi-round LED device with intervals by 3° across 60 LED lamps within a radius of 1.5m in order to determine the angles of subjective edge detection on sound envelopment.

2. Parameters

According to Ando [9], the temporal and spatial parameters of a sound field cover sound pressure level (SPL), first reflection, reverberation time and inter-aural cross-correlation coefficient (IACC) by which the parameters of the three components were set up. Figure 7

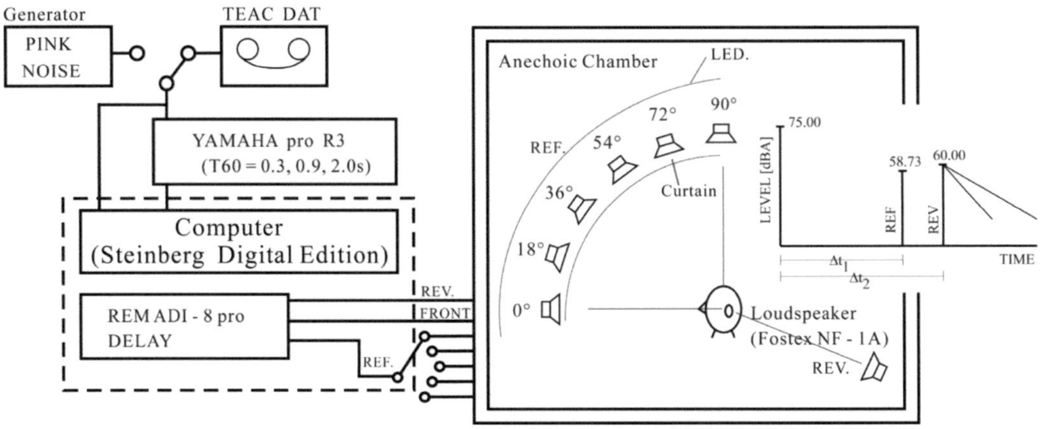

Figure 7. The block diagram of the simulation system for direct sound and two early reflections and the diffused reverberation is attached on the second reflection, which was used in all subjective judgment experiments. Sound pressure levels of the three components were illustrated simultaneously. The direct sound was located in front of subject (ξ=0°) with first reflection at the median plane from η = 18°to 90° and reverberation at clockwise horizontal plane 90° (ξ= 90°, η = 0°).

simultaneously shows the setting up of sound energy in compliance with spatial components of sound energy in a common indoor sound field by the SN ratio of direct sound and first reflection by 15 dB and SPL of direct sound and the other two by 75 dB(A) and 60 dB(A). By the report on the auditory perception in a concert hall by Morimoto [8], reverberation can compose a full image of spaciousness as the second reflection with energy more than the first reflection by 1.27 dB. This is the so-called point of subjective equality (PSE). Thus, the energy of early reflections was reduced to 58.73 dB (SLOW, A weighting, peak). Figure 7 shows the equality. The time gap between direct and first reflection sound (Δt_1) was determined pursuant to research by Morimoto [16] under early reflection sound at 50 (ms) and reverberation at 80 (ms) in compliance with the gap by 1.8 times between early and subsequent reflections by Ando [9]. Also, the author arranged the experiments under RT60 = 0.3s (short), 0.9s (medium) and 2.0s (long) to enhance the impact of reverberation time on spaciousness in a sound field.

3. Determination

- Split judgment (Preliminary)

To prevent image split in a sound field, 36 sound fields randomly comprising of the three Motifs (Motifs A-C with time: 5s) under 3 directions of early reflections (η = 18°, 54°, 90°) and four reverberation times (0.0s, 0.3s, 0.9s, 2.0s) were judged by 15 subjects for 3 times respectively. In this procedure, the subjects confirmed that sound envelopment was perceived as an integrated image without split.

- Edge detection (Primary)

To obtain sound image outline of respective angles, reverberation times and Motifs, 45 sound fields randomly comprising of three Motifs (Motifs A-C with time: 5s) under five directions of early reflections (η = 18°, 36°, 54°, 72°, 90°) and 3 reverberation times (0.3s, 0.9s, 2.0s) were judged by subjects through the sensory threshold of adjustment method for three times respectively. In this procedure, the subjects were asked to answer regarding how the location of the edge of sound envelopment was perceived.

4. Subjects and samples

The subjects of two procedures were 15 male students with normal hearing aged 25±2. In terms of the signal autocorrelation functional theory by Ando [9], a sound source is featured with varied dynamically temporal characteristics critical to spaciousness of a sound field in addition to spectral cues that are called autocorrelation or temporal cues. Table 2 shows details of Motifs A-C.

Source	Title	Composer, writer	Tone	τ_e:ms
Motif A	Royal Pavane	Orlando Gibbons	Andante Downcast	127
Motif B	Sinfonietta, Opus 48; IV movement	Malcolm Arnold	Light Vivid	35
Motif C	Symphony No.102 in B flat major; II movement	Franz J. Haydn	Adagio	65

Table 2. Details of Motifs A-C. Source: BBC (Burd, [8])

3.4. Analyses of subjective source envelope at the horizontal and the median plane

1. Subjective integrity of sound image

The subjective integrity of sound image outline is independent of the angles of first reflection (η = 18°, 54°, 90°) (three-way ANOVA, P = 0.900). Motifs A-C are independent as well (three-way ANOVA, P=0.322). Through the ANOVA, subjective integrity is dependent with the reverberation time (three-way ANOVA, $p < 0.001$) and Table 3 shows the results of a Latin Square Design (LSD) analysis of reverberation times. Results indicate that the subjective integrity of the sound image is not affected by the variation of the reverberation time, but both with and without reverberation time.

Means followed by the same letters are not significantly different at 5% level.			
t Grouping	**Mean**	**N**	**RT60**
A	2.9333	45	0.3
A	2.8667	45	2.0
A	2.8444	45	0.9
B	0.5333	45	---

Table 3. LSD of reverberation times

2. First reflection and edge detection on envelopment

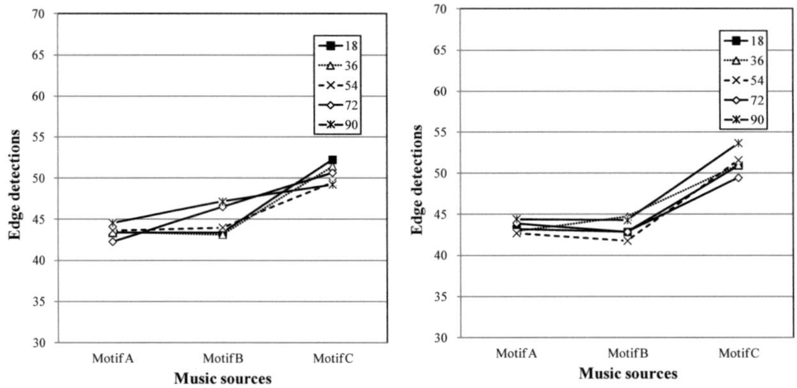

Figure 8. Results of edge detections on Motifs A-C oriented by lateral reflections at the median plane (Left: RT60 = 0.3s ; Right: RT60 = 2.0s)

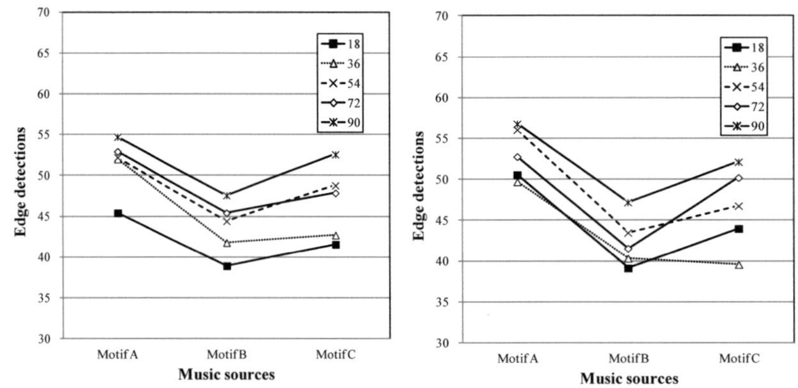

Figure 9. Results of edge detections for Motifs A-C oriented by lateral reflections on the horizontal plane as a reference to Figure 3 (Left: RT60 = 0.3s ; Right: RT60 = 2.0s)

4. Relationship between speech articulation of monosyllable and inter-aural cross-correlation

4.1. An approach on speech intelligibility regarding binaural sensation in a hall

The speech intelligibility for the monosyllables of Chinese in Taiwan area are in agreement with the effective duration of autocorrelation function (τe) of the syllable itself in the same reverberation levels were found (Chen and Chan [21]). On the contrary, it was found (Chen [22]) that they are opposite between speech transmission index (STI proposed by Steeneken

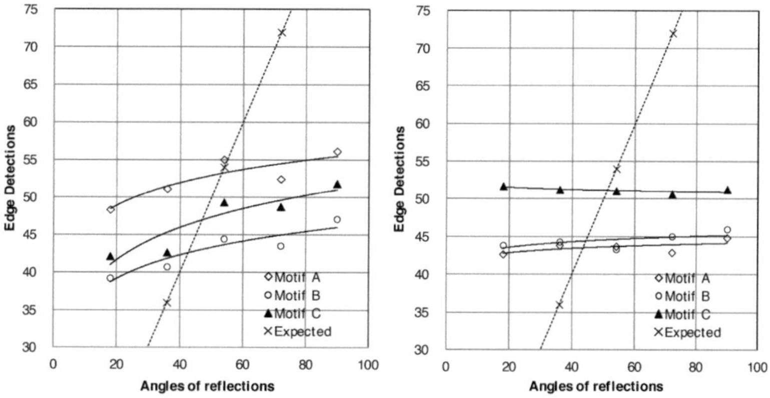

Figure 10. Results of averaged subjective edges values for the significant differences between Motifs A-C oriented by the lateral reflections on the horizontal plane (upper) for mean values at all RT60 conditions, and the source width associated with the τe, ACF of the music sources. However, the source width is independent of the reflections on the median plane (see below).

and Houtgast [23]) and magnitude of inter-aural cross- correlation (IACC) where the slope of ceiling were changed in the hall. However, the range of STI (0.5 ~ 0.7) was quite constricted in this study. Takaoka and et al. [24] once used noises and Japanese language to examine the influence of a sound field's reverberation time and IACC (magnitude of inter-aural cross-correlation function) on speech articulation. It was found that under an IACC condition where SN (signal-to-noise ratio) was between -10dB ~10dB and reverberation time varied between 0.5s ~ 4.0s, no obvious changes were noticed in speech articulation, and that only when SN was lower than -10dB, IACC affect speech articulation within the range of IACC limited in between 0.5 ~ 1.0. Accordingly, this section focuses on a broadened IACC range (0.34 ~ 0.87), and adopted the paired comparison to identify the relationship between speech articulation and IACC with or without reverberate energy in a hall.

4.2. A generalized theory of biaural measurements in a concert hall

1. The IACC of a sound field

In the field of room acoustics, Ando [9] adopted the magnitude of inter-aural cross-correlation function (IACC) to elucidate human ear's spatial impression on sound field, and also determined main diffuse grades and perception of horizontal directionality of acoustic source in a sound field. Tessier and et al., [25] stated that directionality of acoustic source was a physically front-end mechanism of cocktail effect. They researched on voice articulation in noisy environment through acoustic source separation. But the purpose of study would not feed to the systematical hall design. Ando [9] hypothesized that impulse response of each ear on the path of sound transmission was $h_{nl}(t)$ and $h_{nr}(t)$ respectively. Their inter-aural cross-correlation function can represent human's subject sound localization or spatial impression against sound field. The signals *fl* (t) and *fr* (t) of sound's arriving in the ears can serve to express that IACC represents brain's spatial treatment mode, which is defined as follows:

$$\Phi_{lr}(r)=\lim_{T\to\infty}\frac{1}{2T}\int_{-T}^{+T}f_l'(t)f_r'(t+\tau)dt,|\tau|\leq 1ms. \tag{3}$$

Both $f'_l(t)=f_l(t)*S(t)$ and $f'_r(t+\tau_1)=f_{rl}(t)*S(t)$ refer to signals passing through the A-weighting filter which corresponds to hearing perception $S(t)$. Standardized IACC can be modified to Equation (4) from Equation (3) as follows:

$$\varphi_{lr}(\tau)=\frac{\Phi_{lr}(\tau)}{\sqrt{\Phi_{ll}(0)\Phi_{rr}(0)}} \tag{4}$$

Φ_{ll} (0) and Φ_{rr} (0) are monaural autocorrelation functions when delaying τ at the original point (autocorrelation function equals to the average sound intensity of both ears when τ = 0), and total energy arriving both ears is:

$$\sqrt{\Phi_{ll}(0)\Phi_{rr}(0)} \tag{5}$$

However, standardized cross-correlation function in a real room sound field can be modified as follows based on number of reflected sounds and their difference in energy:

$$\varphi_{lr}^{(N)}(\tau)=\frac{\sum_{n=0}^{N}An^2\Phi_{lr}^{(n)}(\tau)}{\sqrt{\sum_{n=0}^{N}An^2\Phi_{ll}^{(n)}(0)\sum_{n=0}^{N}An^2\Phi_{rr}^{(n)}(0)}} \tag{6}$$

where$\Phi_{lr}^{(n)}$ (τ) is the cross-correlation function forming in both ears by the nth reflected sound; Therefore, the grade of inter-aural cross-correlation function can be defined as Equation (7):

$$IACC=\left|\varphi_{lr}^{(N)}(\tau)\right|_{\max} \tag{7}$$

and the maximum delay of signals between both ears is limited to ｜τ｜≦ 1ms.

Moreover, when point source defuses on plane angle ξ(with the front ξ= 0 as datum point) and if the source signal is broadband noise between low and high cut-off frequencies, f_1 and f_2, the inter-aural cross-correlation function can be modified to:

$$\Phi_{lr}(\tau)=H_{lr}\left[\frac{2}{\Delta\omega(\tau-\tau_\xi)}\right]\sin\left[\frac{\Delta\omega(\tau-\tau_\xi)}{2}\right]\cos\left[\frac{\Delta\omega_C(\tau-\tau_\xi)}{2}\right] \tag{8}$$

where H represents power value of each function, τ_ξ represents the left and right delay caused by horizontal angle ξ, and ω is frequency of filter.

where

$$\Delta\omega_C = 2\pi\left(f_2 + f_1\right), \Delta\omega = 2\pi(f_2 - f_1) \tag{9}$$

Figure 11 explains relationship between inter-aural cross-correlation function and various reference factors, while variation width (W_{IACC}) of cross-correlation is as follows when $\frac{\Delta\omega_c}{2}$ is minimal:

$$W_{IACC} \approx \frac{\Delta\omega_c}{4}\cos^{-1}\left(1 - \frac{\delta}{IACC}\right) \tag{10}$$

where δ is the percentage of human ear that can serve to judge change existing in IACC, which is 0.3 normally; Equation (10) shows that maximum W_{IACC} generates the maximum directional perception against acoustic source at horizontal angle ξ. On the contrast, when IACC < 0.15, subjective diffuseness can be perceived.

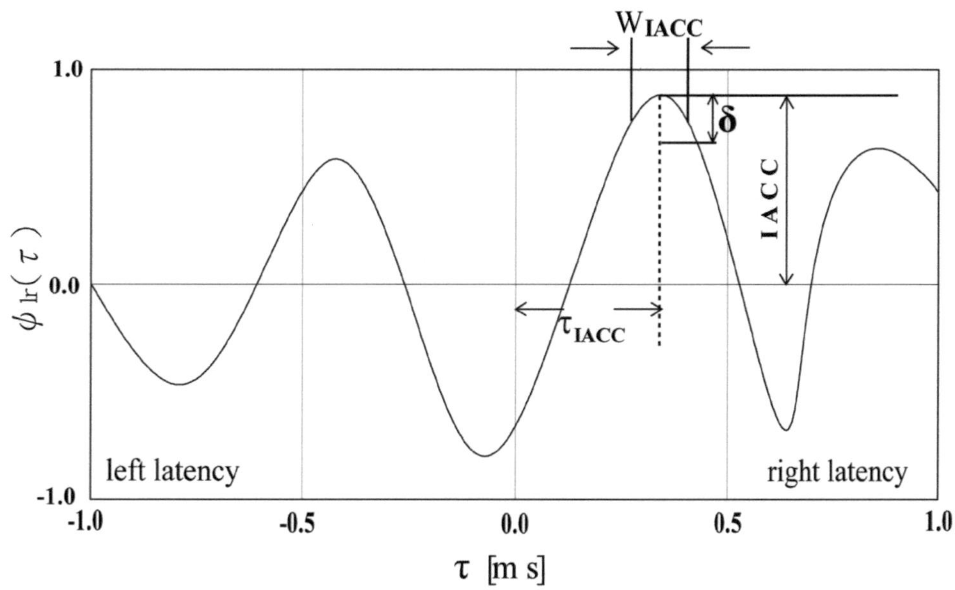

Figure 11. The eigenvalues of standardized IACC can be modified by Equation (4).

Sato, Mori and Ando [26] proposed magnitude of inter-aural cross-correlation function (IACC) and variation width of cross-correlation function can determine magnitude of acoustic sound width (ASW). Since the source used in the experiment was 1/3-octave noise, they found perception of ASW was lessened when center frequency (125Hz – 2kHz) width was enlarged. Therefore, they proposed to define W_{IACC} as a span during which IACC was within 10% of profile scope of cross-correlation figure's maximum, which corresponds to ASW. Schroeder et al [27] found correlation between $IACC_t$ (t = 50 ~ 140ms) and listing preference. Therefore, IACC indeed increases its applicability to subjective diffuse of sound field. As stated in section 2., Chen and Chang [28] used sound field of two reflected sounds to investigate directional perception of subjective source with musical samples, and he found IACC was the dominating factor and inhibited by magnitude of total reflected sound and length of reverberation. Ohnisi and et al., [29] utilized metro station to research transmitting articulation of sound and found that under influence of 1/3-octave background noise, IACC of the diffuse sound field decreased with increase of sound frequency, and articulation of sound transmission was lowered too. Thus property of spatial sound transmission in sound field is related to variation of IACC.

2. Subjective word intelligibility in sound field

As early as the age when telecommunication devices, such as telephone, were first invented, articulation test has been adopted to test perceptibility of auditory sense against language. Such test was employed to test communicating quality between transmitter and receptor. But now, it is applied to test articulation of telecommunication. Licklider and Kryter [30] conducted objective physic and subjective psychological experiments for speech intelligibility (STI) in Bell Telephone Laboratories and Harvard University's Psychological Sound Laboratory respectively in order to establish a set of effective mono-syllabic test lists, known as Harvard P.B.50 word score (Phonetically Balanced Word List, PB). To expel suggestive factors of other speech voice signals during process of measurement from influencing identical accuracy of STI, articulation test lists were composed of a series of common mono-syllables, with each syllable made up of consonant and vowel. Currently, there are many experimental measure methods which adopt this mono-syllabic speech scale in the world such as Diaz and Velazquez's [31] mono-syllabic speech scale for Spanish. Chen and et al., [32] compiled 108 common vocal samples from New Chinese Phonologic Rhymes, which were used in Taiwan area, and summarized six sets of Chinese mono-syllabic subjective speech articulation scale item (hereinafter refer to as "articulate scale") from them. Based on these 6 mono-syllabic sets, this study found reverberation time (RT60) in room less than 1.5 s in the space of the auditoriums, about <12000 m^3, the result of STI was consistent with subjective speech articulation and only varied more obviously in few mono-syllables with nasal or voiceless alveolar affricate consonant. To calculate the ability of speech intelligibility, this study calculated percentage of syllable number the subjects could note down accurately during the test to represent correct answer rate and spatial subjective speech intelligibility.

Morimoto, Sato and Kobayashi [33] proposed interaction between word-intelligibility and word-difficulty, where highly intimate words were used to the perceived test sound. In word-intelligibility, the levels of word recognition were the intelligibility percentage of the test sound released to the subject. The experiment result showed that, word-intelligibility and word-difficulty were extremely negatively correlated. Assuming in a sound field with a higher speech transmission index in a public space, the perception of a word-difficulty was higher than that of word-intelligibility and could be assessed more strictly. When investigating reliability of mono-syllabic speech scale, the issue that Chinese mono-syllables undeniably contains mono-syllables, meaningful and meaningless. This study conducted the subjective psychological experiment by adopting paired-comparison method to solve such vague signals of language expression. By bold assumption that there were only identification method of two-sample which was relatively unaffected by "meaningfulness" and "meaninglessness", so the subjects could easily identify which one was more intelligible. Similarly, Licklider [34] investigated IACC's effect on word-intelligibility under noise masking and found that except the effect of SN ratio, decrease of IACC could improve word-intelligibility in the way that mono-syllables were replaced by short sentences. Chen [35] arranged the recordings of mono-syllables in 7 halls, and found that effect both word-difficulty and word-intelligibility could be separated clearly using accumulated cepstrum of the speech voice.

4.3. Subjective attributes of the sound fields with two initial reflections in relation to mono-syllables intelligibilities

1. Setting and configuration of objective physical quantities

Since the variation range for expanding IACC conditions in the experiment of Takaoka and et al., [24] was too narrow, speakers in semi-anechoic chamber were employed to serve sound field simulation of fewer reflection sound energy from various angles. This system was based on the method of IACC simulation design by Damaske and Ando [36], which allowed individual energy and time delay of direct and reflection sounds in sound field. It was equipped with reverberator to feed subsequent reverberant energy so as to decrease quantity of loudspeakers. This study cited the sound field simulation system in the subjective assessment experiment by Damaske and Ando [36] as reference. In order to simulate different circumstances of room IACC's effect on intelligibility of mono-syllables, this study hypothesized a direct sound in straight front of the subjects, the first and second reflected sounds were hypothesized to transmit to the subjects from different azimuth angles. To further explore the inference by reverberation time of the room, part of the energy of subsequent-RT (RT60) were added to the first and second reflected sounds simultaneously, and then simulated to configure the loudspeakers in the semi-anechoic chamber, whose diagram is shown as Figure 12.

For convenience of the experimental configuration of sound simulated quantity, IACC should be first calculated by adopting Equation (7) from the values of $\Phi_{lr}(\tau)$ and $\Phi_{rr}(\tau)$ measured by Ando [9]. Next, the loudspeakers should be arranged within the range as to generate the IACC in the range of 0.3 to 1.0, where the white noise served as sound source and the dummy head to receive signal. As illustrated in Figure 12, θ_1 and θ_2 were set at 90° and 108° respectively, and with configuration of the IACC measurement was 0.34, 0.56, and 0.87 respectively.

Based on the above simulated configuration, loudspeakers on both sides were added RT energy and set as RT60 = 0.5s and 2.0s respectively. All loudspeakers were 1m from center of the subjects' heads and 1.2m from the ground, while sound pressure was set as 65 dB (SLOW, A weighting, peak) at upper center of the head. Initial reflected sounds mainly simulated the reflection of right and left walls in the simulation of a hall. The delay time and details of sound field are shown as Table 4.

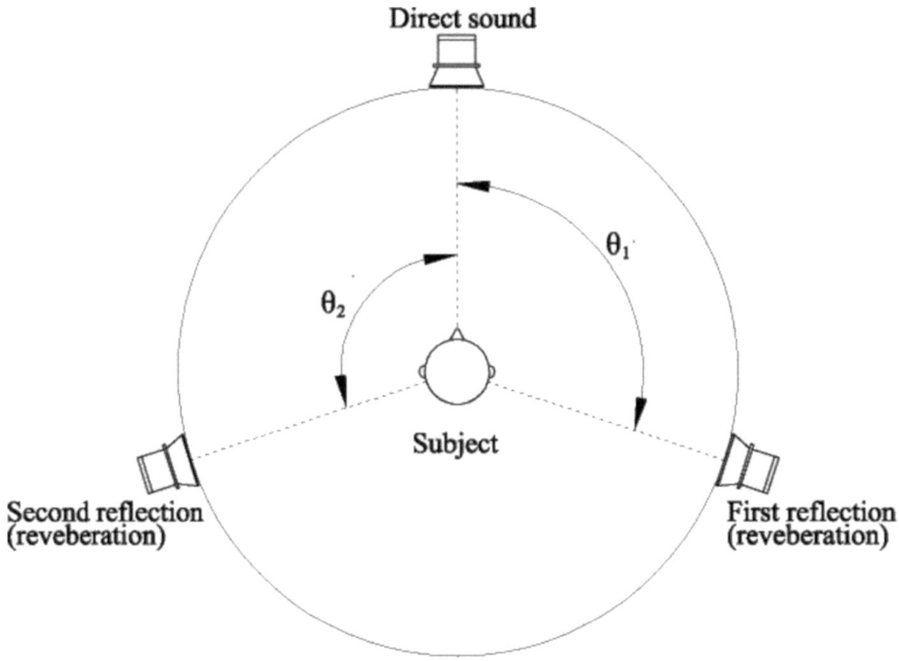

Figure 12. Assumption of IACC configuration was composed by three loudspeakers arranged at different azimuth angles.

2. Sound source

Mono-syllables were same as the research [37] on the correlation between speech intelligibility and continuous brain wave recorded on cerebral cortex, where mono-syllables with higher subjective word-intelligibility such as /heh4/, /ian1/ and /tzuen1/ were figured out, and then compared them with the lower /yu2/.

3. Subjects and experimental method

Total 58 students with average age 23±5 were enrolled as subjects. These subjects were requested to listen and directly answer to experimenter as speech intelligibility. They sat on a fixed chair in the semi-anechoic chamber and concentrated located as Figure 13. The speakers (FOSTEX, NF-1A) were covered with cloth in the semi-anechoic chamber with the light dim. Subjects kept their heads straight ahead and were not allowed to turn, and a repeated test should be avoided in order to avoid over familiarity with the speech samples and thus impairing independence of comparison between sample pairs modified by the assumption of Thurstone's CASE V [38]. This is an obedience to CASE V in paired- comparison theory, that a pair of rivals is independent of each other. In order to quantify the psychological responses of subjective word intelligibility, this study adopted paired-comparison method to gather the scale values of individual syllable, by pairing individual Chinese mono- syllable samples with sound field setting of IACC randomly, and took three different events which had RT60 =0.0 s, 0.5 s, and 2.0 s in turns. Thus each comparison experiment had six samples and 15 pairs, which were treated by different quantified values would be yielded under different IACC and RT60 settings. In distribution of time in psychological experiment, response time from prompting time was 10 s, while interval of prompting between every two samples was 2 s. Each speech dry source had a span about 0.3 s in average, thus time required by every 15 pairs was 3:15 min. Listening test of each speech had 60 pairs. With four speeches completed total 240 pairs of differentiating pairs which were done in four working days.

Items	Conditions
Azimuth angles	Direct (Ch1, 0 deg. straight front to subjects), 1st reflection (Ch2, 90 deg. Ch3,108 deg.), 2nd reflection (Ch4, 90 deg., Ch5, 108 deg.); Added RT energy (Ch2 90 deg., Ch3, 108 deg., Ch4, 90 deg., Ch5, 108 deg.)
Delay gap between the direct and the reflections,	IACC(0.34)- direct : 63.6 dB(A), 1st reflection: 62.7dB(A) delay (9.46ms), 2nd reflection: 62.7 dB(A) delay (17.04ms)
and its SPL setting	IACC(0.56)- direct : 62.8 dB(A), 1st reflection: 60.8dB(A) delay (10.84ms), 2nd reflection: 48.8 dB(A) delay (19.51ms) IACC(0.87)- direct : 64.6 dB(A), 1st reflection: 53.4dB(A) delay (15.48ms), 2nd reflection: 53.4 dB(A) delay (27.87ms)
Reverberation time (RT60)	0.0 s , 0.5 s, 2.0 s
IACC, measured	0.34, 0.56, 0.87

Table 4. Experimental settings

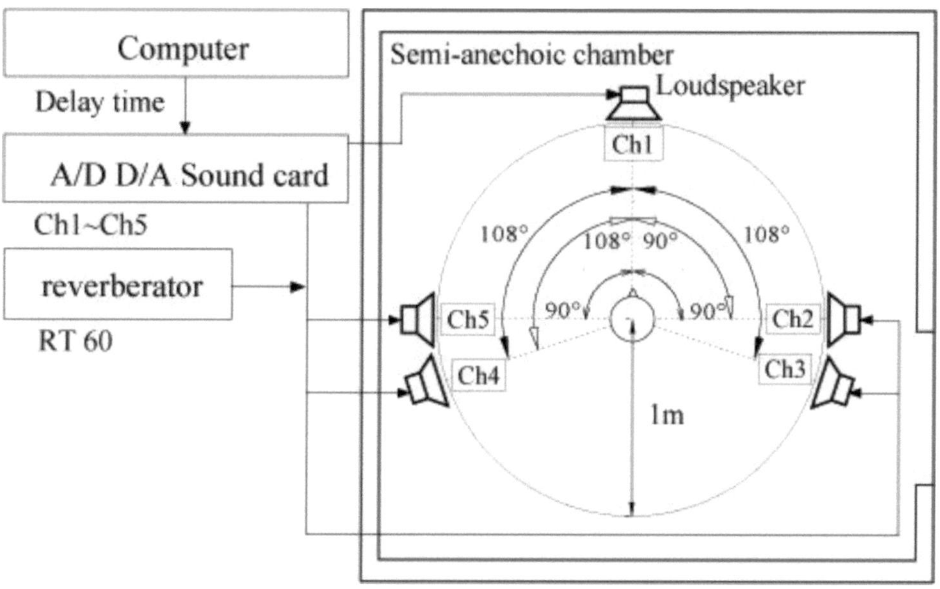

Figure 13. Diagram of experiments

4.4. Analyses of mono-syllabic word-intelligibility

1. The effect of IACC on mono-syllabic word-intelligibility

In order to enhance reliability of the integral answers conducted by paired-comparison method, we counted the numbers of circular-triad once for every subject based on Thurstone's [38] response consistency test for the experiment of every 15 pairs, through which paired - comparison of these 15 pairs were determined effective questionnaires. Subsequently, a test of goodness of fit for comparison quantification model was performed to verify the scale values met the hypothesis of paired-comparison CASE V by Thurstone [38] with respect to effectiveness of difference between stimuli samples and sample size (Mosteller, [39]).

Based on paired-comparison method CASE V by Thurstone [38], average quantified scale value of word-intelligibility of 58 subjects under the conditions of additional RT60 were calculated and shown in Figure 14 ~ 17. Quantified scale value of subjective word intelligibility of mono-syllables under variation of IACC, 0.34, 0.56, and 0.87 showed that trend of subjective higher word-intelligibility before addition RT60 was significant ($p<0.001$).

By ANOVA, the effect of IACC and RT60 on quantified scale values of mono-syllabic subjective word-intelligibility showed that there exist no interaction between these two factors, two-way ANOVA, $F = 0.27$ and $p = 0.90$. But in the case of an individual factor's effect on quantified scale values of mono-syllabic subjective word intelligibility, only RT60 presented significantly, two-way ANOVA, $F = 96.38$ and $p < 0.001$), while the effect of IACC had lower significance, two-way ANOVA, $F = 5.34$ and $p < 0.05$. This result reconfirm that RT60 is independent of IACC in sound field, no matter when with regard to musical preference (Ando [9]) or word-intelligibility.

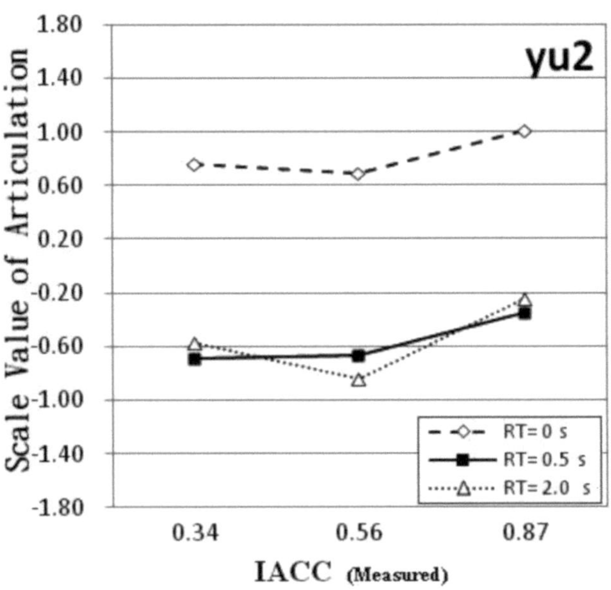

Figure 14. Results of syllable" Yu2"

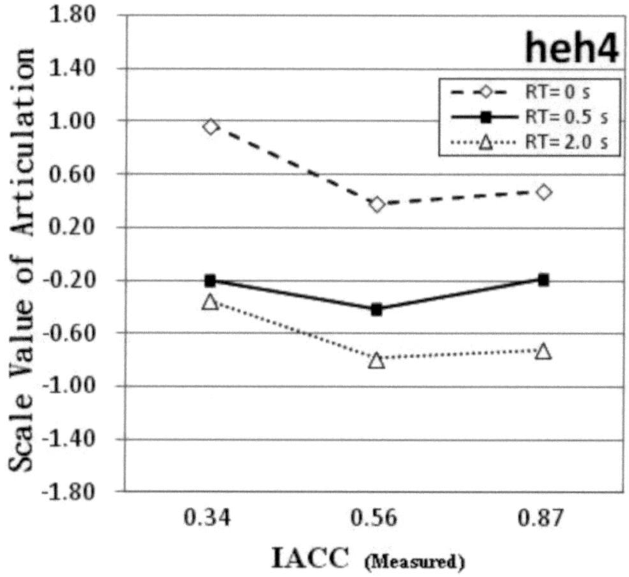

Figure 15. Results of syllable" Heh4"

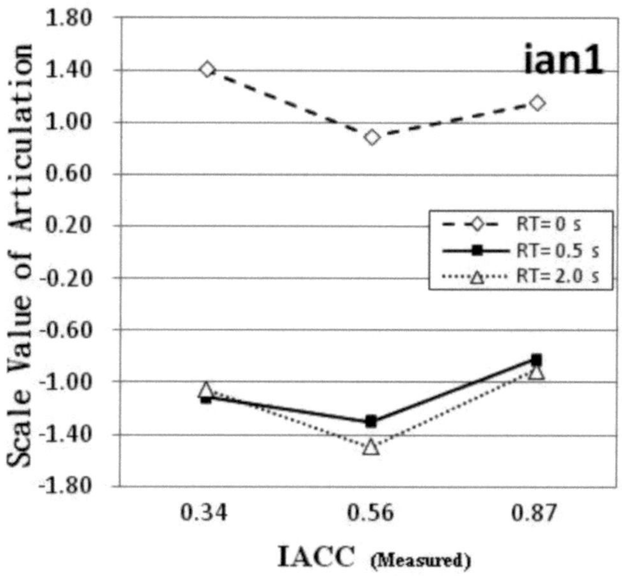

Figure 16. Results of syllable" Ian1"

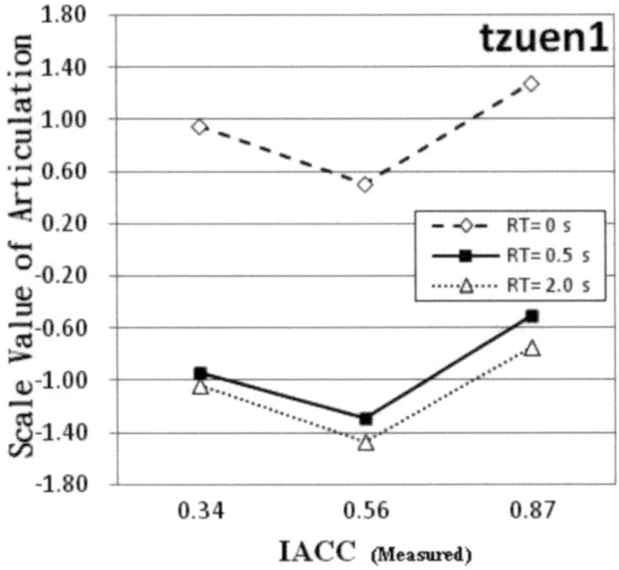

Figure 17. Results of syllable" Tzuen1"

In investigation of the effect of RT60 along on quantified scale values of mono-syllabic subjective word-intelligibility with the setting RT60 = 0.0 s, 0.5 s, and 2.0 s, more significant effect of IACC's variation did not presented. Thus only one-way ANOVA analysis under the environment with RT60 existence and not existence could be performed. The result showed that the effect of IACC's variation was significant in the environment with RT60, by one-way ANOVA $F = 3.74$ and $p < 0.05$. It was doubted of the faith of the results on word-intelligibility is usually changed with regard to IACC in the circumstance of only SN was lower than -10 dB found by Takaoka and et al., [24]. We identify that two reflections of the sound field were not harmful for the word-intelligibility in our settings, and there was no background noise employed here. The setting of RT60 = 0.5 s and 2.0 s adopted here is 1.27 dB in relation to the reflections without reverberant energy at the PSE as stated above (section 2.). Therefore, reflection with RT60 will enhance the variation of IACC on word-intelligibility.

2. The effect of RT60 on quantified scale value of mono-syllabic subjective word-intelligibility

It is clear in Figure 14 ~ 17 that quantified scale values of mono-syllabic subjective word intelligibility obviously changes with RT60. Such change is especially significant between RT60 = 0.0 s and RT60 = 0.5 s. In order to figure out difference among them, this study adopted p value of matrix of Fisher LSD method (Table 5) by multiple mean comparison and found that there was significant difference in quantified scale values of word-intelligibility between RT60 = 0.0 s and RT60 = 0.5 s, $p<0.001$, while there was no significant difference between RT60 = 0.5 s and RT60 = 2.0 s, $p = 0.297 > 0.05$. This result is similar to that of ANOVA on quantified scale values stated as above, suggesting variation between environments of word-intelligibility with and without RT60 was significant. Therefore, Takaoka et al. [24] investigated the cross effect of RT60s in sound field on grades of IACC and found that word-intelligibility between 0.5 s and 4.0 s corresponded with the conclusion that grades of IACC were independent from each other. This study complemented the phenomenon that quantified scale values of subjective word intelligibility was influenced by grades of IACC.

Similarly, by testing p value in the matrix of Fisher LSD method (Table 6) with multiple mean comparison it was clear that there was significant difference between quantified scale values of word-intelligibility of IACC(0.34) and that of IACC(0.56), $p = 0.025 < 0.05$; there was also significant difference between that of IACC(0.56) and that of IACC(0.87), $p = 0.004 < 0.05$; while there was no significant difference between that of IACC(0.34) and that of IACC(0.87), $p = 0.445 > 0.05$. Therefore, it was clear from multiple mean comparison test that the effect of variation in IACC on mono-syllabic word-intelligibility was similar to the variation of musical preference in sound field, which were both related to magnitude of data of standardized IACC grades (Equation (4)). However musical preference was inversely proportional to that and was here inversely proportional to mono-syllabic word-intelligibility, by one-way ANOVA $F = 3.74$ and $p < 0.05$. This finding reconfirms that word-intelligibility under varied IACC is associated with nonlinear response in evaluating the subjective localization of sound sources studied above (Figure 5 of section 2.).

LSD test; variable; Probabilities for Post Hoc Tests. Error: Between MS =.11403, df = 27.00			
RT60	{1} 0.872	{2}-0.706	{3} -0.853
0.0 s	—	0.000*	0.000*
0.5 s	0.000*	—	0.297
2.0 s	0.000*	0.297	—

Table 5. The results of RT60 effect evaluated using p value of matrix of Fisher LSD method

LSD test; variable; Probabilities for Post Hoc Tests. Error: Between MS = .11403, df = 27.00			
IACC	{1} -0.155	{2}-0.481	{3} -0.049
0.34		0.025*	0.445
0.56	0.025*		0.004*
0.87	0.445	0.004*	

Table 6. The results of IACC effect evaluated using p value of matrix of Fisher LSD method

3. Relationship between the parameters within wave's characteristics of IACC and word intelligibility

In order to figure out the correlation between IACC and mono-syllabic word intelligibility in detail, this study used dummy head measurement system to detect parameters which were grades of standardized IACC, delay of inter-aural cross-correlation function (τ_{IACC}), and width of the inter-aural cross-correlation function (W_{IACC}) (Table 7). Sato, Mori and Ando [26] stated in their research that IACC and W_{IACC} could determine acoustic source width (ASW). According to Table 7, the measured data of W_{IACC} in this study was not correlated well to IACC, while τ_{IACC} and IACC showed the opposite trend. Of course, its effect on mono-syllabic word intelligibility also presented RT60 condition under RT60 = 0.5s and 2.0s.

IACC	0.34	0.56	0.87
τ_{IACC}	0.22	0.06	0.09
W_{IACC}	0.19	0.18	0.18

Table 7. The parameters are picked up by wave's characteristic of IACC

5. Conclusions

These facts of section 2. and 3. point out that the temporal characteristics of source signal should be taken into account when estimating and measuring physical measurements, like the lateral energy fraction and the inter- aural cross- correlation coefficient, to estimate source localization sensitivity. For section 4., the experiment of judgment through paired-comparison method, quantified scale values of word-intelligibility was generated based on the hypothesis of CASE V cited by Thurstone [38]. The results show that existence of reverberant energy in a sound field had effect of mono-syllabic word-intelligibility, and that variation of IACC did too. Four mono-syllables with different word-difficulty, subjective mono-syllabic word-intelligibility had certain similar reaction trend under conditions of different IACC and RT60. Results of inductive statistical analyses are shown as follows:

1. As shown in Figure 3, reverberation does not suppress the degree of source directional sensitivity as early reflections after the direct sound, if their ratios of lateral to frontal sound energy are the same. Even though music source directional concept of auditory distinction is inverse to spaciousness of a sound field. The spaciousness is not at all suppressed by levels of early reflections at the PSE at echo threshold for all levels of reverberation whenever the reverberation (RT60) was fixed at 0.3 or 0.9 s concluded by Morimoto [1] as well.

2. As shown in Figure 4, the source directional sensitivity caused by different source signals is suppressed by τe of ACF of itself even if the sound field includes both early reflections and reverberation and with their preferred initial time gap after direct sound signals. This finding is an important problem with which to perceive the localization of performers for assisting visual enjoyment in concert halls. The temporal structure of source signal to auditory spaciousness is first discussed out of sound energy or directional mentioned before.

3. The source directional sensitivity are quicker as the coming direction of early reflection sounds located at the azimuth angle from -36° to -54° (Figure 5) as the early reflection functions as lateral energy fraction in a simulated diffuse sound field. The sound incidence angle of -54° is found upon the deep notch and peak at 54° of the curve in the transfer function of the ear canal entrance in a free sound field, especially in the frequency range from 2 to 4 kHz (Mehrgardt and Mellert [7]). It is obvious that source localization at a horizontal plane angle is dependent upon the transfer function of the ear canal.

4. As shown in Figure 7, with a fixed gap between the sound pressure levels of the three spatial components, direct sound, first reflection and subsequent reverberation, the reverberation discerned will affect the capability of an integrated image envelopment without split, demonstrating that reverberation is crucial factor to the envelopment perceived but the edge judgment of image boundary is not affected by reverberation time (Figure 7). This finding is in harmony with the result of sensitivities on reflective signal localization researched in section 2. The reverberation does not suppress the orientation of both source image edges and reflection incidences in addition to the perception of source image split.

5. As shown in Figure 8, the first reflection from the upper hemisphere at the angles $\eta = 18°$, 36°, 54°, 72°, 90° does not affect the edge judgment of image boundary for music Motifs A-C. The ability of edge localization is independent with the angles of first reflection in median plane but sound source. Rakerd, Hartmann and McCaskey [19] that found listeners failed to identify noises with roved the location when the spectral structure was at a high frequency because the spectral structure was confused with the spectral variations caused by different location. Such is the fact that music with temporal variation leads to confusion regarding the edge of the sound image with a reflection incidence on the median plane in a diffuse sound field. Morimoto and Nomachi [11] have both explained that localization accuracies of sound images on the median plane produced by both binaural disparity cues and frequency cues. Morimoto, Yairi, Iida and Itoh [20] concluded when the source is a wide-band signal, only higher frequency components (> 2 kHz) are dominant on the median plane localization. However, they did not consider that a source with a wide-band sound in temporal variation provides the changing of the source width conception during a concert. Thus, it is presumably difficult to account for the different locations on the median plane of a music source in a hall except for during a recital of an instrument with a higher frequency tones.

6. As shown in Figure 9 and Figure 10, the difference of Motifs and the subjective judgment of edge detections of sound image outline on horizontal plane are interdependent, and the tempo of music proposed by Ando [9] are related well. This evidences that the temporal cues are important to the subjective edge determination and source localization.

7. Depending on one-way ANOVA for the environment with and without reverberation, the result of word intelligibility showed that variation of IACC (0.34 ~ 0.87) had significant effect on the environment with reverberation (0.5s ~ 2.0s), $F=3.74$ and $p<0.05$. Takaoka and et al., [24] reported that IACC influences on speech articulation within the range of 0.5~1.0 only when SN was lower than -10dB under RT60 = 0.5s ~ 4.0s. There is no conflict between

these two results because word-intelligibility was not affected by RT60 varied from 0.5s to 2.0s in our research when reverberation was constantly 1.27 dB higher than the reflections. Reflections with RT60 enhance the variation of IACC on word-intelligibility at the PSE of equal spatial impression in the source width. They have obviously confirmed evidence by similar W_{IACC} of varied IACC's environments in Table 7, which may indecate the source width of sound signal stated above.

8. Figures 14 ~ 17 illustrate the interaction between RT60 and mono-syllabic word articulation, which show that IACC's effect on mono-syllabic word- intelligibility significantly varied with span of RT60 ($p<0.001$ ANOVA).

9. Test on matrix of Fisher LSD with multiple mean comparison confirmed in Table 5 showed that quantified psychological scale values of word-intelligibility were significantly different between RT60 = 0.0 s and RT60 = 0.5 s, $p < 0.001$, while not significantly different between RT60 = 0.5 s and RT60 = 2.0 s, $p = 0.297 > 0.05$. This finding indicates that the source signal image was buried by reverberation and would defect word-intelligibility such as source split as induced by with or without reverberation as investigated in section 2. Similarly, Table 6 confirmed that quantified psychological scale values of word-intelligibility were significantly different at IACC(0.34) and IACC(0.56), with $p = 0.025 < 0.05$, was significantly different at IACC(0.56) and IACC(0.87) too, with $p=0.004 < 0.05$, while was not significantly different at IACC(0.34) and IACC(0.87), with $p=0.445 > 0.05$. The nonlinear responses in evaluating word-intelligibility, source edge and localization of spatial impression at the horizontal plane under varied IACC are presumably influenced by transfer functions of the ear canal entrance as measured by Mehrgardt and Mellert [7].

Glossary of symbols

ASW	apparent source width
IACC	inter-aural cross-correlation
τ_{IACC}	inter-aural time delay at cross-correlation function
ICF	inter-aural cross-correlation function
W_{IACC}	inter-aural variative width at cross-correlation function
LL	listening level
RT60	reverberation time
τe	effective delay of autocorrelation function
ACF	autocorrelation function
PSE	point of subjective equality

http://dx.doi.org/10.5772/56616

SRP	stationary random processing
DAT	digital auditory tape cassette
ϕlr	binaural normalized cross- correlation function
$\Phi rr(\tau)$	mono- aural autocorrelation function
$\Phi lr(\tau)$	binaural cross correlation function
η	vertical angles at an median plane, 0° started from the front of head at ear height
ξ	angles at clockwise horizontal plane, 0° started from the front of head at ear height
LEV	listener envelopment
SPL	sound pressure level
SN	logarithm of signal over noise energy, denotes by decibel
Δt_1	delay gap between direct and first reflection in a defuse sound field
LSD	Latin Square Design
STI	speech transmission index
δ	percentage at the peak of wave form in inter-aural cross-correlation function, as the definition of W_{IACC}
$IACC_t$	time gap of sound signal in inter-aural cross-correlation
PB	Phonetically Balanced Word List
/yu2/	example of a mono-syllable in Taiwanese's life speech

Author details

Chiung Yao Chen*

Architecture Department, Chaoyang University of Technology, Taiwan

References

[1] Morimoto M, Posselt C. "Contribution of reverberation to auditory spaciousness in concert halls [J]," J Acoust Soc Jpn, (E)10, 2: 1989, 87-92.

[2] Ando Y, "Calculation of subjective preference at each seat in a concert hall [J]," J Acoust Soc Am, 74: 1983, 873-887.

[3] Barron M, Marshall A H. "Spatial impression due to early lateral resection in concert hall: the deviation of a physical measure [J]," J Sound and Vibration, 77 (2): 1981, 211-232.

[4] Inoue T, Nishi T, Wakuri T, Shimizu Y, Kawakami F. "Relation between the lateral component of early reflections sound energy and the spatial impression," Proc Spring Meet Acoust Soc Jpn, 1987: 557-558 (in Japanese).

[5] Hasegawa H, Takehashi K, Ayama M, Kasuga M. "Effects of visual information on sound image localization [J]," J. Image info and Tele Eng, 55, 3: 2001, 455-462 (in Japanese).

[6] Marple S. L. "A new autoregressive spectrum analysis algorithm," IEEE Trans. Acoust., Speech, Signal Process, ASSP. 28, 1980, 441-454.

[7] Mehrgardt S. and Mellert V., "Transformation characteristics of the external human ear," J. Acoust. Soc. Am., 61, 1977, 1567-1576.

[8] Burd V. A. N. "Nachhallfreie Musik fur Akustische Modelluntersuchungen," Rundfunktech. Mitteilungen, 13, (1969) 200-201.

[9] Ando Y. "Concert Hall Acoustics," Berlin Heidelberg, New York, 1985.

[10] Chen C. Y. "Effects of reverberation time and sound source characteristic to auditory localization in an indoor sound field," ICSV Cairns Australia, 2007, 9-12.

[11] Morimoto M. and Nomachi K., "Binaural disparity cues in median-plane localization."J. Acoust. Soc. Jpn. (E) 3,2, 1982, 99-103.

[12] Morimoto M. and Aokata H., "Localization cues of sound sources in the upper hemisphere."J. Acoust. Soc. Jpn.(E) 5,3, 1984, 165-173.

[13] Morimoto M., Yoshimura K., Kazhiro I. and Motokuni I., "The role of low frequency components in median plane localization." J. Acoust. Soc. Jpn.(E) 24, 2, 2003, 76-82.

[14] Morimoto M. and Iida K. "Relation between auditory source width and the law of the first wave front." J Acoust Soc Jpn, 49, 2: 1993, 43-55 (in Japanese).

[15] Barron M, Marshall A H. "Spatial impression due to early lateral resection in concert hall: the deviation of a physical measure." J Sound and Vibration,,77 (2), 1981, 211-232.

[16] Morimoto M. and Posselt C. "Contribution of reverberation to auditory spaciousness in concert halls", J. Acoust. Soc., 1989, 87-91.

[17] Asahi N. and Matsuoka S. "Effect of the sound anti resonance by pinna on median plane localization - Localization of sound signal passed dip filter-," Tech. Rep. Hear. Acoust. Soc. Jpn., 1977, H-40-1.

[18] Sato S. and Ando Y. "Apparent source width (ASW) of complex noises in relation to the interaural cross-correlation," Journal of Temporal in Architecture and the Environment, 2002, 29-32.

[19] Rakerd B., Hartmann W. M. and McCaskey T. L. "Identification and localization of sound sources in the median sagittal plane," J. Acoust. Soc., 106 (5), 1999, 2812-2820.

[20] Morimoto M, Yairi M, Iida K, Itoh M, "The role of low frequency components in median plane localization [J]," J Acoust Sci & Tech. 24, 2: 2003, 76-82.

[21] Chen, C. Y. and Chan M. H., "A Study of the Chinese Speech Intelligibility in Halls in Relation to the Autocorrelation Function Model: The Case of Chinese in Taiwan," J. Archi., Architectural Institute of the R. O. C, 57, 2006, 55-68.

[22] Chen, C. Y. "Investigation on speech intelligibility in respective to the variation of diffusing panels in a hall," 16th Proceeding of the Archi. Institute of the R. O. C, E66, 2004, 336 - 341. (in Chinese)

[23] Steeneken H. J. M. and Houtgast T. "The Modulation Transfer Function in Room Coustics as A Predictor of Speech Intelligibility, " Acoustica 28, 1973, 66-73.

[24] Takaoka T., Morimoto M., Sato, H. and Semba Y. "Effects of inter-aural cross-correlation of speech intelligibility and background noise on listening difficulty," J. Acoust. Soc. Jpn., NO. 9, 63, 2007, 520-528 (in Japanese).

[25] Tessier E., Berthommier F., Glotin H. and Choi S. "A case front- end using the localization cue for segregation and then cocktail-party speech recognition," Proc. IEEE Int. Conference on speech Process (ICsP), Seoul (1999).

[26] Sato S., Mori Y. and Ando Y. "The subjective evaluation of source location on the stage by listeners," In: Music and concert hall acoustics. Y. Ando and D. Noson (eds.). Academic Press, London, 1997, 117-123.

[27] Schroeder M. R., Gottlob D., and Siebrasse K.F. "Comparative study of European concert halls: correlation of subjective preference with geometric and acoustic parameters," *J. Acoust. Soc. Am*, 56: 1974, 1195-1201.

[28] Chen C. Y. and Chang Y. R. "A Study of Test Method for Absorption Coefficient of Material through Cross-Correlation in an Anechoic Chamber- The porous plane materials as example", Journal of Technology, 20 (2005). (in Chinese)

[29] Ohnisi Y., Maeda K., Morimoto M. and Sato H."Acoustic characteristics of background noise at subway stations,"Proc. WESPAC IX (2006).

[30] Licklider J. C. R. and Kryter K. D. "Articulation tests of standard and modified interphones conducted during flight at 5000 and 35,000 feet (OSRD Report 1976)", Cambridge, MA: Harvard University, Psycho-Acoustic Laboratory (1944).

[31] Diaz C. and Velazquez C. "A Live Evaluation of the RASTI -Method", Applied Acoustics, 46 1995, 363-372.

[32] Chen C. Y., Chen L. S., and Lin W., "A Study on Evaluation Method of Chinese Articulation Standard of Speech Intelligibility for Sound Field in Taiwan", *Journal of Architecture,* No.43, Architectural Institute of the Republic of China, 2002, 27-36. (in Chinese)

[33] Morimoto M., Sato, H. and Kobayashi M. "Listening difficulty as a subjective measure for evaluation of speech transmission performance in public spaces," *J. Acoust. Soc. Am.*, 116, 2004, 1607-1613.

[34] Licklider J. C. R., "The influence of inter-aural phase relations upon the masking of speech by white noise," J. *Acoust. Soc. Am.* 20, 1948, 150-159.

[35] Chen C. Y., "Syllables Intelligibility in Relation to the Autocorrelation and Cepstrum Model: The Case of Chinese in Taiwan," POMA Volume 15, pp. 015002 (June 2012); Acoustical Society of America, ISSN 1939-800X (online).

[36] Damaske P. and Ando Y. "Interaural cross-correlation for multichannel loudspeaker reproduction," Acoustica, 27, 1972, 232-238.

[37] Chen C. Y. and Chen U. S., "Cortical continuous brain waves in relation to the speech intelligibility of mono-syllables in Taiwan," master dissertation, Graduated School of Architecture and Urban Planning, Chaoyang Univ. of Tech., Taichung, Taiwan (2010).(in Chinese)

[38] Thurstone L. L. "A Law of Comparative Judgment", Psychol. Rev., 34, 1927, 273-289.

[39] Mosteller F., "Remarks on method of paired comparison: I. The least squares solution assuming equal standard deviation and equal correlation," Psychometrika, 16, 1951, 3-9.